GLACIER SCIENCE
AN INTRODUCTION

GLACIER SCIENCE
AN INTRODUCTION

Prof. Avinash Tiwari

Delve
PUBLISHING

www.delvepublishing.com

Glacier science:An introduction

Prof. Avinash Tiwari

Delve Publishing
4164 Lakeshore Road
Burlington, ON L7L 1A4
Canada
www.delvepublishing.com
Email: orders@arclereducation.com

ISBN: 978-1-77956-356-9 (Hardcover)

ABOUT THE AUTHOR

Prof. Avinash Tiwari, Vice-Chancellor of Jiwaji University, Gwalior, brings a wealth of academic and administrative experience spanning over 33 years. Prof. Tiwari has held pivotal roles in higher education, including as Professor, Associate Professor, and Assistant Professor. His academic qualifications include NET and a Ph.D., which have underpinned his distinguished career dedicated to teaching, research, and institutional administration. With over 31 years of administrative leadership, he has served as the Head of Department, Examination Controller, Dean of Faculty, and member of UGC and NAAC national bodies. Since his appointment as Vice-Chancellor in December 2021, Prof. Tiwari has steered Jiwaji University to unprecedented achievements, including securing the NAAC A++ Grade, UGC Category I recognition, and a ₹ 100 crore grant under the PM-USHA program. His efforts have significantly elevated the university's academic and infrastructural standing while fostering innovation through MOUs with 35 national and international organizations. Under his visionary leadership, Prof. Tiwari has implemented numerous transformative initiatives. These include the establishment of a Medical College and Research Center, the promotion of e-governance, and the enhancement of sustainability efforts within the campus. A staunch advocate for educational accessibility, he has championed initiatives such as the "Bandi Se Bandhu" scheme for educating prisoners and programs to empower tribal communities. Additionally, his specialization in Food Technology, Agro Ecology, and Tourism Management, complemented by his international academic engagements, reflects his commitment to advancing global education and research. Recognized with honors like "Gwalior Ratan" and leadership awards, Prof. Tiwari exemplifies excellence and dedication in the field of higher education.

TABLE OF CONTENTS

LIST OF FIGURES

PREFACE

The majority of individuals who reside in temperate regions typically only come into contact with limited amounts of ice on cold days or as small chunks in drinks. In its common form, ice seems to be a hard, brittle, and typically slippery substance, yet in glaciers and ice sheets, it may take on a vast range of unexpected and captivating behaviors. It can to mold itself around hills and valleys, creep and slide from high snowfields down to lower terrain. Huge troughs can be formed, and the entire landscape can be changed by the ice scrubbing dirt and materials or covering significant areas in glacial deposits. The climate of Earth can be changed by glaciers and ice sheets, which can lower temperatures in the oceans and atmosphere and have a substantial impact on the hydrological cycle. In addition, they hold invaluable archives of previous climate change trapped in ice crystals and air bubbles, and the fluctuations of glacier boundaries serve as clear indications of current climate change. Moreover, glaciers can retain and abruptly release enormous amounts of meltwater, resulting in disastrous. Massive amounts of glaciogenic material are constantly being dug out of densely populated areas and used for the construction of roads and buildings. The mountains of waste from homes and businesses are constantly filling in the pits that are left behind. Thus, a significant amount of the population is impacted by glaciers and glaciation, either directly or indirectly.

This book introduces the various aspects of glaciers and their associated processes and can be used by students and researchers in the field.

– Author

INTRODUCTION

Glaciers are an essential component of the global cryosphere. Although they make up less than 1% of the total amount of ice that is trapped on land as sheets and glaciers, they are currently the main cryospheric factor contributing to the rise in global sea levels. A few thousand square kilometers are made up entirely of glaciers and the larger ice caps that extend beyond the edges of the major ice sheets in Greenland and Antarctica. Because of this enormous scale, it is implied that remote sensing tools, whether they are used from satellite platforms or the air are crucial to the scientific study of glaciers and their mass balance. The majority of glaciers, whether found in the Arctic or the alpine regions are receding. Providing quantitative datasets that can be used to gauge the speeds at which these changes are occurring is undoubtedly one of the main goals of remote sensing. To detect any increase in the rate of mass loss from glaciers on our warming planet, these data must be collected over the time series and throughout the largest geographic area. During glacial cycles, ice sheets actively adapt to changes in climate by changing their extent and spatial structure. The sediments and landforms left behind after they melt serve as an important record. This implies that no location on Earth is too remote to research past glacial situations, given there is a vast corpus of material and a variety of techniques available. Growing precipitation, floods, and increased demand for clean, fresh groundwater are all consequences of climate change, raising questions about our comprehension of the characteristics of the glacial sediments that cover large portions of the continents.

CHAPTER 1

INTRODUCTION TO GLACIERS

Contents

1.1 INTRODUCTION

The glaciers can have a significant impact on both human communities and natural ecosystems (Hodson et al., 2008). The hydrosphere, atmosphere, biosphere, and lithosphere are the other four main parts of the Earth system and the cryosphere exchanges mass and energy continuously. Because they are always expanding and contracting in response to variations in temperature, precipitation, and other variables, glaciers are considered sensitive. Sea level rise, increasing risks from avalanches and flash floods, and disappearing water supplies have all become pressing issues in the last few decades as a result of the ice volume decreasing. Large landscapes carved by glaciers have generated massive deposits of glacial debris, rock, and silt (Cook & Swift, 2012).

Figure 1.1 Ice sheets in Greenland

Source: https://commons.wikimedia.org/wiki/File:East_Greenland_Ice_Sheets_%285563148850%29.jpg,Creative Commons License.

While ice sheets in Greenland (Figure1.1), Antarctica (Figure 1.2), and smaller glaciers and ice caps offer rich archives of past environmental conditions, moraines, which are set down like tidemarks at former glacier boundaries, provide important sources of information on past glacier activity and climate change.

Figure 1.2 Antarctica

Source: https://commons.wikimedia.org/wiki/File:Antarctica-cold-floating-48178.jpg,Creative Commons License.

We can think of glaciers as systems having inputs, outputs, and interactions with other systems, such as the atmosphere, oceans, rivers, and the terrain, to comprehend how they behave or respond. The mass of glaciers comes into the system as rock fragments and snowfall. The mass has potential energy because it is elevated in the gravitational field; this energy is used up as the glacier moves downslope. Ice must be warmed or melted using energy, which must then be released from the system as heat or water. Potential energy is transformed into work, pushing rock and ice into the sea. The term 'mass balance' refers to the ice that is deposited and lost in glacier systems. The flow of the glaciers carries snow and ice until it reaches places where it is lost to the system due to melting, evaporation, or the breaking of ice blocks or icebergs, a phenomenon known as ablation. When inputs exceed losses due to climatic and topographic factors i.e., when accumulation surpasses ablation, glaciers expand; conversely, when outputs surpass inputs, glaciers retreat. Although the mass of the glacier stays constant, the overall energy store within it might fluctuate over time due to energy exchanges between the glacier and the atmosphere.

1.2 GLACIER SYSTEMS

One of the most crucial elements of glacier systems is meltwater (G. H. Brown, 2002). Water from melting glaciers gets into the oceans and the atmosphere, among other aspects of the global hydrological system, and has a significant impact on the hydrology of proglacial areas. Meltwater carves the landscape as it flows from the edges of glaciers to the ocean, leaving deposits of silt, sand, gravel as well as ravines. Liquid water within and beneath glaciers and ice sheets has a significant impact on glacier behavior, regulating flow rates and affecting the rates and processes of erosion and deposition. Glacier flow moves snow and ice from areas of accumulation to areas of ablation. Processes like sliding, ice deformation, and glacier bed deformation, are responsible for flow. Glaciers travel across the terrain through one or more of these processes, transporting snow and ice to places where ablation exceeds accumulation and allowing meltwater and icebergs to escape the system. The balance between driving forces i.e., gravitational acceleration and resisting forces i.e., drag at the glacier's bed and borders determines the rates and patterns of glacier motion. The rates of glacier flow must match the rates at which snow and ice are deposited and removed in the accumulation and ablation areas when the entire system is in balance. However, a number of factors can upset the system equilibrium, causing glaciers and ice sheets to behave in an unpredictable manner.

Water is always traveling between the seas, atmosphere, cryosphere, and hydrosphere in its liquid, solid, or gaseous state (Peixoto, 1995). Rarely do these fluxes balance, which causes storage to fluctuate. Sea level can be directly impacted by glacier fluctuations since less water in the oceans corresponds to more storage in glaciers and ice sheets. Variations in ice volume also affect the distribution of mass on the planet and the loading patterns on the crust, which further affects regional and global sea levels. They are excellent carriers of detritus, moving massive volumes of silt, sand, gravel, and boulders from their original locations across distances of several hundred kilometers. This debris is left over in a variety of environments, from the ocean floor to the surface of glaciers. The variety of sediments and landforms known as corries, drumlins, and moraines that were left behind after deglaciation are the remnants of glaciers and ice sheets in the mid- and high-latitude regions. The extent and behavior of historical ice masses can be reconstructed using these sediments and landforms, which also offer crucial hints about the functioning of the global climate system both now and in the past. From an engineering perspective, glacial sediments and landforms are especially significant since a great deal of human activity occurs in areas impacted by glacier erosion or deposition.

Climate and terrain influence the shape that glaciers take, and each glacier's morphology is specific to its location on the surface. Because of this, there is a wide range of glacier morphologies, ranging from the smallest niche glacier to the ice sheet. The terrain is submerged, at least in parts, by ice sheets and ice caps, and the main patterns of ice flow are mostly unaffected by bed inclinations. On the other hand, faster-moving ice streams and outlet glaciers are typically found in troughs in the outer areas of ice sheets and ice caps (Figure 1.3).

Figure 1.3 Ice caps

Source: https://commons.wikimedia.org/wiki/File:Renland_Ice_Cap.jpg, Creative Commons License.

The large ice masses in Nordaustlandet, Ellesmere Island, Baffin Island, and Iceland are referred to as ice caps, whereas the ice masses currently covering the majority of Antarctica and Greenland are called ice sheets (Dowdeswell et al., 1997). Ice domes, which are elevated regions of comparatively slow-moving ice, and ice streams and outlet glaciers, which are zones of faster flow through which the majority of the ice is discharged towards the periphery, are divisions of ice sheets and ice caps. A wide, raised portion of an ice sheet or ice cap is called an ice dome. The underlying land surface can be a large topographic low, like the Laurentide and Scandinavian Ice Sheets, or a topographic high, like East Antarctica. In ice-sheet domes, ice thicknesses can frequently reach over 3000 meters, whereas in ice-cap domes, the thicknesses are just a few hundred meters (Demorest, 1943). A central ice dome or group of coalescent domes evacuates mass radially outward, serving as dispersal centers. Although there is no horizontal velocity component, vertical flow nevertheless occurs as a result of ice compaction and deformation. Ice sheets typically have several splits and are rarely simple domes. As an ice sheet grows and deteriorates, ice splits will shift in location. As a result, there will be significant changes in the patterns of ice flow, and ice-directional indicators such as flutings and drumlins will overlap or possibly destroy one another.

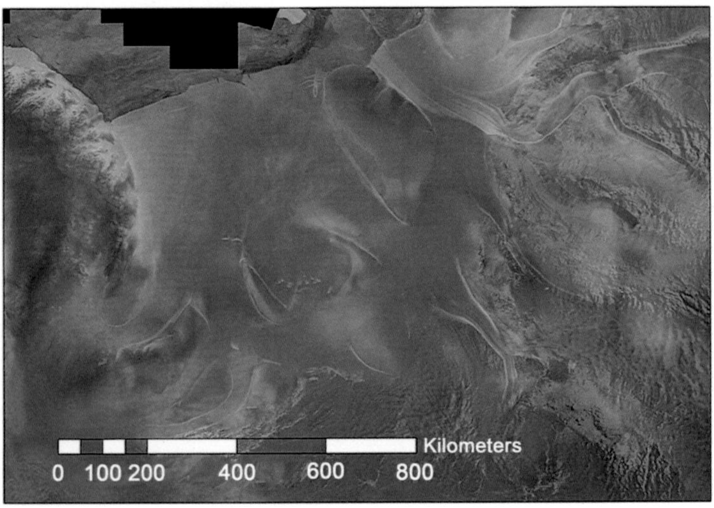

Figure 1.4 *Rutford Ice Stream*

Source: https://en.wikipedia.org/wiki/Rutford_Ice_Stream#/media/File:FRicestreams.jpg, Creative Commons License.

Sheet flow, which takes place inside ice dome areas, and stream flow, which happens in outlet glaciers and ice streams, are the two categories into which the movement of ice within ice sheets and ice caps can be demarcated. Channeled ice that shoots out from the inside of ice sheets and caps in the form of outlet glaciers and ice streams, moves quickly. While ice streams are surrounded by slowly moving ice, outlet glaciers are found in troughs or valleys. An example is the Rutford Ice Stream in Antarctica (Figure 1.4), which is bordered on one side by the Ellsworth Mountains and on the other by a section of the ice sheet that moves more slowly, demonstrates properties of both outlet glaciers and ice streams (Alley & Bindschadler, 2001). The 700 km Lambert Glacier in Antarctica, which is the longest glacier in the world, and the Jakobshavn glacier, which is one of the fastest glaciers in the world, are both made of ice streams in some places and outlet glaciers in others. The existence of highly crevassed zones along their borders allows ice streams to be distinguished from the surrounding ice. Deep troughs can form when there are depressions in the glacier bed that are frequently occupied by outlet glaciers and ice streams. Compared to ice domes, outflow glaciers often have low steep gradients. The bulk of ice sheet discharge is caused by swift movement through outlet glaciers and ice streams, which also contribute to the majority of iceberg debris that ends up in the oceans. Because of their dynamic properties and significance in draining ice sheets, ice streams and outlet glaciers have a significant impact on the behavior and stability of ice sheets. As a result, they have recently attracted a lot of attention from researchers. Unlike ice caps, ice fields have a flow that is impacted by the underlying topography and lack a simple, dome-like cover. Anywhere there is enough altitude for ice buildup and generally soft but locally stressed terrain, ice fields will form. Such ice masses include the Patagonian Ice Fields, the Tien Shan/Kunlun Shan ice fields in China, the St Elias Mountains ice fields in the Canadian Yukon Territory/Alaska, and the Columbia Icefields in the Canadian Rocky Mountains. All of these ice fields are drained by major valley glaciers.

Figure 1.5 Valley glacier

Source: https://commons.wikimedia.org/wiki/File:Franz_Josef_Glacier_View_Into_the_Valley.jpg, Creative Commons License.

1.3 VALLEY GLACIER

A valley glacier (Figure 1.5) is created if ice flows into a deep bedrock valley from an ice field or cirque (Bennett et al., 1999). These glaciers may create dendritic networks resembling those found in fluvial systems, or they may have a single branch. Similar to rivers, valley glaciers can be grouped based on where they are in the hierarchy of drainage basins. The structure and lithology of the bedrock frequently have a substantial impact on the form of valley glacier networks. Valley glaciers can have very wide altitudinal ranges and frequently have steep bedrock slopes underneath them. The existence of ice-free slopes dominating the glacier surface is the major characteristic that sets valley glaciers apart. These slopes are significant contributors of avalanches, which build up snow and ice, and of rock fragments that fall onto the surface of glaciers. Transection glaciers are typically formed by ice cover in areas of deeply cut mountain landscapes that have been glaciated. In these conditions, a system of radiating valleys is formed by glaciers flowing down from multiple directions, yet the ice run-offs the preexisting drainage. With ice diffluences, where ice flow divides and sends branches down two or more channels, and confluences, where ice flow from two or more tributaries converges to form a single unit, intersection complexes create patterns resembling networks. The armchair-shaped bedrock hollow that serves as an accumulation basin, particularly for snow pushed by wind, determines the morphology of a cirque glacier.

Figure 1.6 *Cirque glaciers*

Source: https://commons.m.wikimedia.org/wiki/File:Glacier_Gebroulaz_vu_du_Cirque_du_Rosoire_%286%29.jpg, Creative Commons License.

Cirque glaciers (Figure 1.6) come in a range of shapes and sizes, from those that are fully contained inside the bedrock hollows that they support to those that form the larger valley glaciers in areas that have had more glaciation. Piedmont glaciers are formed wherever valley glaciers emerge onto lowland regions after passing through bedrock troughs. Large portions of piedmont glaciers, such as Skeiðarárjoŝkull in Iceland and Alaska's Malaspina Glacier, lie below the equilibrium line altitude. The 600m thick Malaspina Glacier fills a large depression that is up to 250 m below sea level (Telbisz et al., 2019). It is completely contained within the ablation zone. In the Canadian High Arctic, there are a few smaller instances of subpolar glaciers breaking away from plateau icefields and landing in the lowlands of U-shaped valleys.

Thin layers of snow and ice that cling to the slopes of mountains are known as ice aprons, the smallest glacier ice volumes. Glacierets are thin ice patches that form in depressions on less steep terrain caused by avalanches and snow fall. A glacier is referred to as a niche glacier when its location is determined by a rock in the exterior sides of a mountain or valley. Because of internal deformation or sliding, niche glaciers and ice aprons move fast. When the internal resistance is overcome by the forces exerted by the snow weight and surface gradient, glacier movement begins. Some researchers estimated that this barrier will be achieved when snow reaches between 30 and 70 meters from the base(Benn et al., 2012). When a glacier accelerates into an icefall, it produces an area of high flow where the ice stretches under high tensile pressures, creating many crevasses and shattering the ice surface into unstable ice pieces called seracs. A zone of compressive flow is produced at the foot of an icefall when the flow of ice slows down, sealing crevasses and thickening the ice. There are some areas of exposed granite cliffs between the top of the icefall and the glacier below. Ice cover is not always continuous through icefalls. When seracs collapse, ice flows through the icefall in the form of sporadic avalanches that build up cones of crushed blocks at the base.

1.4 GLACIER TONGUES

Low-gradient, floating glacier tongues (Figure 1.7) known as ice shelves are created when glaciers move offshore or where sea ice develops due to surface and bottom accretion. Ice shelves are found only in polar regions; they are the largest along the Antarctica coast. Ellesmere Island, northern Greenland, and the Antarctic Peninsula are surrounded by a smaller ice shelf (Dowdeswell & Jeffries, 2017). Ice rises occur anywhere floating ice shelves get thick enough to ground on offshore shoals or coastal shelves. These may develop as a result of surface accumulation-induced local thickening, ice flowing over the obstruction, or a combination of the two. An ice rise that originates locally can have its own radial flow pattern, distinct from the surrounding ice shelf flow direction. Nearly 16 million km^2 of the Earth's surface are currently covered with glacier ice, the majority of which is found inside the Greenland and Antarctic Ice Sheets (Frezzotti & Orombelli, 2014). The remaining 3% is made up of glaciers, ice fields, and ice caps that are found worldwide in hilly or high-latitude areas. These are mostly found in the northern hemisphere, more precisely in the region surrounding the Arctic Ocean basin, in mountainous coastal areas such as Alaska and Norway, and in high mountainous continental regions such as the Alps and the Himalayas. Significant glacier ice cover can be found in the southern hemisphere in the Andes, New Zealand, and sub-Antarctic islands like South Georgia. The interaction of temperature, topography, and effective precipitation is reflected in the distribution of glaciers. Global variations occur in these characteristics, particularly with latitude, altitude, and distance from a water source.

Figure 1.7 Glacier tongue

Source: https://commons.wikimedia.org/wiki/File:Thwaites_Glacier_Tongue.jpg,Creative Commons License.

Because there is less energy available to melt snow and ice at high latitudes due to the low solar angle, glaciers will be closer to the poles. Since mean temperatures drop with air density, the chance of glaciers surviving improves with altitude for any given latitude. The global pattern of glaciation is produced by the interaction of latitudinal and altitudinal elements. This pattern shows that glaciers only occur in high-altitude equatorial regions and occupy lower elevations as one moves toward the poles. For instance, glaciers can occur at sea level in high-latitude areas like the Canadian and Greenland High Arctic, but they can only be found above 4500–5000 m in the tropical mountains of Ecuador, Irian Jaya, and central Africa (Hastenrath, 2009). Glacier equilibrium line altitudes (ELAs) ascend from both poles in the direction of the equator, peaking in the subtropical Mexican Highlands and Andes of Bolivia. Variations in precipitation complicate this process, towards the equator, resulting in lower glacier ELAs in regions with heavy snowfall. This is true for the Patagonian Ice fields, where snowfall can approach seven meters annually, and the highly glacierized southern side of the Alaska range. Similarly, compared to arid regions like Bolivia and Mexico, ELAs are comparatively lower in Ecuador, which is a part of the Intertropical Convergence Zone. Aspect, relief, and the glaciers' distance from water sources affect their existence. Mountain shape, aspect, and continentality as well as size must be considered.

Figure 1.8 *Himalayas*

Source: https://commons.wikimedia.org/wiki/File:Himalayan_Horizon_Aerial_Ladakh_India.jpg, Creative Commons License.

Mountain landscapes like the Himalayas (Figure 1.8), the European Alps, the North American Rockies and Coast Ranges, the South American Andes and the Southern Alps of New Zealand, and are large enough and at a sufficient altitude to harbor significant glacier ice cover. At small scales, slope is important for determining the placement of glaciers, where the regional snow line is situated just below mountain peaks. Due to the varying absorption of solar radiation and precipitation in high-relief terrain, snow line elevations can fluctuate by several hundred meters. Because north-facing slopes receive the least amount of solar energy and are protected from the dominant south-westerly winds, they operate as snow traps and have the lowest snowlines in many northern hemisphere ranges.

Some peaks frequently stay snow-free even in the most glaciated regions, even while nearby highland plateaus at lower elevations harbor large ice fields or ice caps. This is a function of relief, which includes altitude and summit shape. In general, small summits with steep sides won't have enough room for snow accumulation, while higher mountains will be able to store enough snow to produce a glacier. Avalanches can carry snow to lower elevations when mountain slopes are too steep to store it, occasionally feeding glaciers much beyond the regional ELA(Anderson et al., 2018).

Habitat might be sufficiently cold to support glaciers, yet it might also be located far from the closest source of precipitation. Despite having exceptionally low winter temperatures, many continental interior regions receive very little precipitation. The amount of precipitation in these regions could not be enough to sustain even mountain ice fields, therefore glacialization might depend on how close a location is to the water source. The distribution of glacier ice in the Canadian Arctic is a good illustration of this; large ice fields on south-east Ellesmere Island and Devon Island are fed by the 'open north water' of Baffin Bay, but owing to moisture decline, ice thicknesses and glacier cover decrease in a westward direction.

Similar to this, Antarctica is very dry, and most ice sheet accumulation happens near the shore. The amount of glacier ice on Earth is continually fluctuating, ranging in duration from a few days to millions of years. Approximately 10% of the Earth's surface is currently covered by glaciers and ice sheets, although this percentage has fluctuated dramatically throughout the planet's history(Allison et al., 2021). When the major Quaternary glaciations peaked, for instance, there was more than three times as much ice as there is now, but for extended periods, there was significantly less. Although the potential effects of glacier retreat have drawn a lot of attention in recent decades, the current glacier decline is merely the most recent development in a long history of constant change. Understanding the size of previous glacier fluctuations is crucial for putting current changes in glacier extent into context.

1.5 EARTH AND GLACIERS

The Earth has seen multiple cycles over hundreds of millions of years, with intervals of high average glacier ice cover, or icehouse conditions, and intervals of low or no ice cover, or greenhouse conditions. These long-term cycles are with shorter-term variations as glaciers continuously modify their volume, extent, and flow rates in response to changing environmental settings. From a geological perspective, the Earth is presently covered with massive ice sheets in the polar regions and extensive mountain glaciers. Since the start of the Late Cenozoic, approximately 34 million years ago, ice sheets and glaciers have never completely vanished(Davies et al., 2012).

The Earth went through an extended greenhouse period before the Late Cenozoic, during which glacier ice was either completely absent or extremely limited. Before then, there were significant glaciers from the Late Devonian to Early Carboniferous, the Permo-Carboniferous, and the Late Ordovician. When portions of the supercontinent Gondwana occupied the south polar regions, all of these glaciations took place.

Average global temperatures were well below freezing, the global hydrological cycle was greatly reduced, and glacier ice extended into equatorial latitudes during the Makganyene glaciation in the Palaeoproterozoic and the Cryogenian period in the Neoproterozoic. Icehouse occurrences have garnered a lot of attention because they were linked to important shifts in the evolution of life and atmospheric oxygen levels. The Earth's average surface temperature is influenced by a wide range of causes. The quantity of shortwave radiation that reaches Earth has gradually increased in billion-year periods due to a gradual increase in solar radiation.

1.6 CARBON DIOXIDE

It is a notable fact that the Earth's climate system has maintained within the boundaries necessary to support life despite this significant, long-term drop in solar output. It is now commonly acknowledged that greenhouse gas concentrations in the atmosphere, which have a significant impact on the near-surface energy balance, regulate the global climate. Tectonic processes have had a significant impact on the atmospheric content of carbon dioxide (CO_2) hundreds of millions of years ago (Müller & Dutkiewicz, 2018). Weathering of silicates and subsequent deposition in oceanic sediments remove carbon from the atmosphere. At convergent plate edges, sedimentary rocks can then be subducted, and volcanic outgassing (Figure 1.9) eventually releases CO_2 back into the atmosphere after a prolonged period of storage inside the lithosphere.

Figure 1.9 Volcanic outgassing

Source: https://commons.wikimedia.org/wiki/File:Vulkan_La_Palma_2022_NACH_Ende_des_Ausbruchs_%28Annamarie_Ursula%29_P1400249.JPG,Creative Commons License.

Thus, the balance between carbon dioxide release from volcanoes and removal through weathering is reflected in atmospheric concentrations of carbon dioxide. Variations in carbon intake and release caused by biological (including anthropogenic) activities and other processes are included in the geologic carbon cycle. According to certain theories, the presence of a supercontinent and low levels of volcanic activity favor low atmospheric CO_2 levels and icehouse conditions, whereas continental drift and many dispersed continents favor high CO_2 greenhouse conditions. While the extended Mesozoic to Early Cenozoic heat interval coincided with CO_2 levels generally greater than 1000 ppm, the Permo-Carboniferous and Late Cenozoic glaciations were linked to consistently low atmospheric CO_2 levels (~500 ppm) (Goddéris et al., 2014). On the other hand, it appears that the short-lived Ordovician glacier occurred at a time when atmospheric CO_2 levels were more than ten times higher than they are now. Large icehouse–greenhouse temperature shifts in the Proterozoic have also been attributed to tectonic forces. Some have argued that the Cryogenian icehouse formed as a result of CO_2 drawdown after the disintegration of the Rodinia supercontinent. The earth may have been locked in a "Snowball Earth" condition during the peak of glaciation when significant losses of shortwave radiation from highly reflecting ice surfaces kept temperatures below freezing on a global scale. It may have taken millions of years for volcanic outgassing to build up atmospheric CO2 to escape this extreme icehouse. However, after the ice melted and shortwave radiation losses decreased, heat feedback was set off, creating an extreme greenhouse environment. Many modifications and alternatives have been put up in response to the intense controversy around this paradigm. For a considerable amount of time, geochemists and Earth system modelers may be occupied with solving the mystery of the vast Proterozoic glaciations in a thorough way and in line with all available geological data.

The location of continents, mountain ranges, and channels, which limit patterns of atmospheric and oceanic circulation, as well as the processes by which energy is transferred around the planet, also have a significant impact on climate. For instance, it has been suggested that the uplift of the Tibetan Plateau, the Himalayas, and the western Cordillera of North and South America, which enhanced CO2 drawdown by weathering, was the primary cause of the beginning of glaciation in the Mid-Cenozoic. The Drake Passage between the Antarctic Peninsula and South America deepened, which promoted the establishment of the Antarctic Circumpolar Current and decreased heat transfer from lower latitudes, contributing to cooling in Antarctica. Shorter-term periodic oscillations between colder temperatures when glaciers and ice sheets are larger and warmer temperatures when ice cover is significantly decreased are observed over icehouse periods. The Late Cenozoic is the period for which the timing of these oscillations is better understood. The oxygen isotope ratios in marine sediments and other techniques can be used to compute the global ice volumes. A standard chronology for the Quaternary period, roughly the last two million years, up to and including the present is also provided by the marine oxygen isotope record. Marine isotope stages (MIS) are numbered in reverse chronological order, starting from the present and going backward with even numbers for cold periods and odd numbers for warm times(Pages, 2016). MIS-1 designates the current warm period, known as the Holocene, and MIS-2 designates the final epoch of fully glacial conditions. Regardless of the location or amount of ice sheets, the climate varies in a quasiperiodic manner across all glacial periods. Since the early Oligocene, when ice sheet glaciation began, the obliquity cycle has been the dominating wavelength. Numerous explanations have been put out to explain this shift in the periodicity of global climate cycles, the majority of which center on the possibility that nonlinear Earth system processes could enhance eccentricity forcing. There has been a recent argument that the 23,000-year precession cycle is reflected in the 100,000-year climate cycle, with four to five precessional cycles linked to major glacial periods(Imbrie et al., 1993). This theory holds that eccentricity is not the main motor, but rather that it modulates precession forces to pace the system.

1.7 ICE SHEETS

Up until the Middle Miocene climatic optimum about 14–10 Ma, the Antarctic Ice Sheet rose and fell multiple times(Levy et al., 2022). After that, it remained stable, but it was still prone to fluctuations. The ice sheet covered an area of more than 15,106 km^2 at the Last Glacial Maximum (LGM), which occurred 20,000 years ago(Siebe et al., 2017). It reached nearly to the edge of the continental shelf. The northern hemisphere's glaciation began about 3.2 million years ago(Willeit et al., 2015). The Laurentide Ice Sheet was the largest of the ice sheets in the northern hemisphere, covering a substantial portion of northern North America at its greatest extent. The Laurentide Ice Sheet meets the Cordilleran Ice Sheet in the west and the Innuitian Ice Sheet in the north at the glacial maximum; however, at times of reduced ice extent, the latter two functioned as separate entities. The Cordilleran Ice Sheet covered the Puget Lowlands in Washington State and the westmost point of the Aleutian Islands during the Last Glacial Maximum. The Canadian Arctic Archipelago was once covered by the Innuitian Ice Sheet, which was divided between an eastern alpine region and a western lowland part. The ice sheets were emptied by many paleo ice streams. Unlike the Laurentide Ice Sheet, which was able to last through some interglacial periods, the Greenland Ice Sheet was able to spread over the adjacent continental shelf during glacial periods. There is evidence that the Greenland Ice Sheet was smaller than it is today during the last interglacial, the Eemian, which ended 125,000 years ago(Reeh, 1991). In glacier accumulation zones, annual layers of snow and ice deposit provide a stratigraphic record of previous environmental conditions. Trapped air bubbles are samples of the atmosphere during the time of ice formation, and the isotopic composition of the water molecules that make up the ice allows one to reconstruct past air temperatures. Furthermore, radioactive fallout, wind-blown dust, and volcanic tephra are examples of impurities found in glacier ice that provide high-resolution archives of the composition of the snow as it changes. Ice cores must be dated in some way in order to offer valuable records of environmental change. Annual accumulation layers are typically visible in the higher levels of cores as alternating bands of foams/bubbles, transparent ice that may be dated by counting down from the surface, much like tree rings. Moreover, annual layers that indicate seasonal changes in air temperature or solute loading can be recognised using chemical or isotopic signatures. Individual layers could not be visible in deeper, older core sections, in which case dating is accomplished using alternative methods.

Since the 1970s, extensive deep-drilling projects in the Antarctic and Greenland ice sheets have produced incredibly significant records of environmental change in both polar regions through the analysis of ice cores. The Dye 3 ice core, measuring a total of 2037 meters in length, was drilled by the Greenland Ice Sheet Project (GISP) between 1979 and 1981 (Dansgaard et al., 1985). GISP2 produced the world's deepest recovered ice core at the time, measuring 3053.44 meters in length (Mayewski & Bender, 1995). The

Greenland Ice Core Project (GRIP) was a multidisciplinary European research endeavor organized by the European Science Foundation. GRIP successfully drilled a 3028 m ice core to the Greenland Ice Sheet's bed between 1989 and 1992 (Souchez, 1997). The North Greenland Ice Core Project (NGRIP) established a drill site. In 2003, bedrock was reached, 3085 meters below the ice surface. Part of the previous (Eemian) interglacial epoch was included in the continuous environmental record of the last 123,000 years that the NGRIP core gave(Andersen et al., 2006). Antarctic ice cores indicate the ice sheet's high stability and slower rates of accretion. The remarkable 740,000 years of the European Project for Ice Coring in Antarctica (EPICA) core from Antarctica, cover the last eight complete glacial cycles (Wolff et al., 2010). Shorter useful records have been recovered from several additional Antarctic locations, such as the West Antarctic Ice Sheet Divide (WAIS Divide) Ice Core Drilling Project and the Japanese Dome Fuji Station (Dome F). Additionally, ice core records have been collected from high-altitude sites in Asia, South America, and Africa as well as from the ice caps on Arctic islands. Some of the drilling locations are over 6000 meters above sea level. Below the surface of a glacier, air bubbles become trapped and isolated when snow turns into ice. The bubbles represent fossil air, which preserves atmospheric samples from the moment they were sealed off. Thus, it is possible to track changes in atmospheric chemistry over many thousands of years by analyzing the composition of these bubbles in ice cores. Depending on the ice temperature and accumulation rate, trapped air in polar ice may be several hundred to several millennia younger than the surrounding ice due to the isolation of bubbles, also known as gas occlusion, which usually happens at depths of 50–150 m. When drawing conclusions regarding the causal relationship between atmospheric chemistry and paleoclimatic substitutions found in ice cores, care must be taken due to uncertainties around the age difference.

1.8 GREENHOUSE GASES

Due to their ability to regulate the earth's temperature, greenhouse gases including carbon dioxide (CO_2) and methane (CH_4) have received special attention in recent years. Using a deep core drilled at Vostok, Antarctica, variations in CO_2 and CH_4 have been determined for the last eight glacial–interglacial cycles. The concentration of greenhouse gases fluctuates remarkably, with peak concentrations occurring during warm periods and low concentrations denoting cold spells. These changes reflect variations in the climate. The correlation found between greenhouse gas concentrations and climate change is due to feedback loops involving glacial, oceanic, atmospheric, and biological systems. Specifically, minor variations in climate lead to adjustments in the amount of trace gases produced and stored by marine and terrestrial vegetation, ultimately causing additional climate shifts.

Figure 1.10 *Chernobyl reactor accident*

Source: https://commons.wikimedia.org/wiki/File:Chernobylreactor_1. jpg,Creative Commons License.

The Vostok record also demonstrates that the present CO_2 and CH_4 concentrations (1,700 ppb and 360 ppm, respectively) are the highest in at least the last 700,000 years (Ruddiman & Raymo, 2003). Thus, the ice core record offers a crucial viewpoint on the ongoing discussion over anthropogenic greenhouse gas emissions as well as the patterns and mechanisms behind climatic change. Glacier ice impurities can also serve as archives of historical environmental conditions. Tephra horizons and acidity peaks record the deposition of aerosols inserted into the atmosphere by volcanic eruptions. Windblown dust and soluble ions can yield data on the strength and direction of atmospheric circulation.

1.9 MARKERS

Acidity peaks are a result of sulfur dioxide produced by large eruptions and their atmospheric conversion to sulphuric acid aerosol. The majority of the acidity peaks correspond to past eruptions, such as Tambora and Krakatoa in the East Indies. The peak connected to the Laki Fissure Eruption in Iceland in 1783, which resulted in 10,000 fatalities there and increased mortality throughout Europe, is especially notable(Sonnek et al., 2017). Radioactive fallout from atmospheric nuclear bomb testing conducted in the 1950s and early 1960s and the 1986 Chornobyl reactor accident (Figure 1.10) is present in ice cores all over the world (Wakeford et al., 2010).

In addition to serving as a warning against the imprudence of exploding nuclear weapons, this fallout offers helpful stratigraphic markers. Researchers conducted a reconstruction of the mass balance of the Kongsvegen/Sveabreen glacier complex in Svalbard, utilizing radioactive fallout from the Chornobyl reactor catastrophe in 1986 and Russian nuclear testing in 1961–1962. Through the melting of snow or ice at the glacier's surface, bed, or inside the ice, water can get into glacial drainage channels. Furthermore, precipitation, subsurface runoff, and groundwater are examples of extra glacial sources of water that can reach glaciers directly. Surface meltwater production exhibits significant spatial and temporal variability contingent on local climate conditions and other variables. Melting can happen every day of the year in the ablation zones of tropical glaciers. Most glaciers undergo annual cycles of melting, in addition to shorter-term fluctuations brought on by shifting meteorological conditions. Seasonal changes in water output are mostly unaffected by melting that occurs within or beneath glaciers in response to frictional heat produced by ice deformation or sliding. This melting occurs over a considerably longer period of time. Large fluctuations in basal and internal melting, however, could occur over extended periods. Water can flow sub-glacially, along or beneath the bed (subglacial flow channel), over glacier surfaces (supraglacial), or through the ice (englacial flow path). While water can move through glacier systems very quickly, with less lag time between the formation of meltwater and runoff, water can also be held on, in, or beneath glaciers for variable lengths of time. After refreezing, storage can be done as ice or as a liquid. Because of this, runoff from glaciers varies greatly in both space and time, based on factors such as fluctuations in the quantity of storage, the nature and effectiveness of transport pathways, and variations in meltwater output. The exact path that water will take over, though, or beneath a glacier is determined by the interaction of two factors: (1) hydraulic potential gradients that propel water flow; and (2) flow resistance. The characteristics and size of glacial drainage routes can quickly change in response to shifting conditions because ice can melt, refreeze, and distort. Surface channels or conduits in or under the ice can expand or contract over time depending on the equilibrium between melting, freezing, and ice deformation.

1.10 ENERGY EXCHANGES

Energy exchanges between the water and the surrounding ice cause the ice to melt or the water to freeze. Water flowing through a channel or passage produces frictional heat because the energy needed to overcome fluid viscosity is lost. A portion of this energy is used to heat the water, but energy losses from melting ice will balance this out. There exists an equilibrium water temperature for which the generation of heat is precisely balanced by the melting of the wall. This holds for any given flow conditions. Water temperature in channels with sediment floors is greater than expected because the silt layer prevents heat transfer between the ice and the water. Because snow and ice have various permeabilities, supraglacial drainage systems on snow-covered surfaces and bare ice have quite different characteristics. Water easily seeps underneath snow surfaces as melting starts because of the linked pore holes. Percolating water will refreeze and create layered ice if the freezing point is reached in deeper portions of the snow. Latent heat is released during this process, progressively warming the snowpack. Refreezing will stop when the snowpack reaches the melting point(Zhou et al., 2008). Subsequently, melted water gathers inside the snow. Firn aquifers can retain vast amounts of water and can reach several meters in thickness, which can greatly delay the runoff from glaciers. Large swamps can develop where hydraulic gradients are minimal, particularly in regions with low surface slopes, making traveling on glaciers challenging and uncomfortable. In freshwater aquifers, integrated drainage systems may take a while to form and runoff may be delayed for a few days or weeks. When the water content rises, rills and surface channels form after the water first inclines to drain laterally via the snow/ice network of linked pores. As drainage improves over time, water is released from storage and the glacier's runoff increases quickly. Since undamaged glacier ice is nearly impermeable, meltwater flows off the surface rather than percolating downward. Compared to channels with a rock or sediment floor, ice-walled channels typically have larger flow velocities due to their minimal roughness. Supraglacial channels must have channel downcutting rates greater than the rate at which the nearby ice surfaces are ablated. Deep, permanent channels can occur if subaerial melt rates are low, as they are on debris-covered glaciers or in cold regions. In streams with sediment floors, meandering occurs when the outside of bends erodes and the inside of bends is deposited. In supraglacial channels with ice walls, there is no comparable depositional process; instead, meanders develop by erosion. Surface melting gradually removes the abandoned upper portions of channels, and meander amplitude tends to rise over time. Structures like crevasses and foliation can also greatly influence surface drainage patterns; in certain cases, this results in the formation of channel networks with strong linear elements.

1.11 MODELS AND STUDIES

Predictive numerical models have been developed because glacier hydrological systems are crucial in controlling proglacial runoff and subglacial pressure conditions (Arnold et al., 1998). Compared to the early analytical studies, the models that are currently available are significantly more sophisticated; however, assumptions still need to be made to represent some significant physical processes. From the surface energy balance, the model determines the amount of meltwater produced. It then determines how water is transported over and through the glacier and into the outflow stream. The basal drainage network topology and the locations where surface meltwater reaches the bed were deduced using known moulin positions. Subglacial water pressure history and diurnal, seasonal, and intermediate-scale oscillations were among the key elements of the observed proglacial hydrographs that the model successfully replicated. To accommodate flow, the subglacial sheet is allowed to extend vertically, and as the water sheet thickens, so does its hydraulic conductivity. Water can flow through or be stored within the supraglacial and englacial drainage systems, contingent on recharge rates and antecedent circumstances, because the subglacial sheet and the underlying aquifer are connected to these systems.

CHAPTER 2

GLACIAL GEOLOGY

Contents

2.1 INTRODUCTION

The study of landforms and sediments produced by glaciers and ice sheets is known as glacial geology. Ice sheets and glaciers have expanded and melted many times throughout history, making them an essential component of our environment. The ice sheets and glaciers that formed and receded during the Cenozoic Ice Age are responsible for the current topography in many mid-latitude regions. The last 65 million years, or the Cenozoic, have seen significant changes in our climate. Ice sheets in Greenland and other parts of the Arctic formed after the Antarctic Ice Sheet (Ehlers & Gibbard, 2007). Large mid-latitude ice sheets later formed in Patagonia, New Zealand, Europe, Scandinavia, and North America. The glacial landforms and sediments that these ice sheets generated are cues of their passage, having drastically altered the terrain underneath them. This record demonstrates that these ice sheets have not only grown and melted regularly over the past two million years due to variations in the global temperature but have also contributed to climate change through interactions and modification of the atmosphere. Knowing these ice sheets and glaciers is essential to comprehending the dynamics underlying global climate change. The glaciers of the Cenozoic Ice Age left behind a unique environment made up of many distinct landforms and deposits in many places on the planet. Even now, this glacial landscape is still present. These glaciers also provide the aesthetically pleasing glacial environment that may be found in many highland regions of North America and Europe.

2.2 PALAEOGLACIOLOGY

These glacier traces can be used to recreate them and learn more about their behavior through the sediments and landforms. Palaeoglaciology is a topic that is becoming important as we try to figure out how the glacial system interacts with the rest of the global system. We may analyze the growth, degradation, and climate interactions of glaciers by reassembling the glaciers that formed the glacial landscapes and investigating their formation. Although the current period is optimistically referred to as the Postglacial, there is no reason to assume that large glaciers or ice sheets will not return to the mid-latitudes in the future. As a result, we can start making predictions about what will happen when the mid-latitude ice sheets return. The greatest ice sheet on Earth is the Antarctic Ice Sheet, which encases 98% of the continent. The ice sheet is nearly 4 km thick when it covers deep subglacial basins, despite having an average thickness of about 1.6 km. With the larger East Antarctic Ice Sheet on one side and the smaller West Antarctic Ice Sheet on the other, it is split in two by the Transantarctic Mountains. Nearly 90% of the glacial ice, or around 70% of the freshwater on Earth, is found in this continent (Khilchevskyi &

Karamushka, 2020). Global sea levels would rise by roughly 60 meters if the ice sheet disintegrated. Rainfall rates are generally quite low in the center of the continent, often as low as 20 mm annually; but, along the coast, where the air is more humid, precipitation rates increase. Because of the potential for it to collapse or rapidly disintegrate, the West Antarctic Ice Sheet (Figure 2.1) has drawn the greatest interest from scientists.

Figure 2.1 West Antarctic Ice Sheet

Source: https://commons.wikimedia.org/wiki/File:.00_1128_Schelfeiskannte_in_der_Antarktis.jpg, Creative Commons License.

The ice sheet rests on top of a basin whose mean elevation is lower than the sea, which could be unstable. Sea levels might rise by as much as 6 meters for several centuries if the West Antarctic Ice Sheet were to break down. The Polar plateau has less moisture, snowfall is estimated in millimeters to centimeters every year. Recent observations of the ice sheet peripheral retreating have led to suggestions that inner thickening on the Polar plateau balances this retreating. However, it is hard to determine with any degree of certainty because comprehensive mass balance studies of the ice sheet are lacking. Because of this, we are unsure of the ice sheet mass balance, whether it is positive or negative. The longest continuous deep ice cores, which cover eight glacial cycles over the previous 740,000 years, have been drilled on the Polar Plateau (Elzinga, 2017). These cores contain data on the climate and atmospheric history of Antarctica.

2.3 ICE STREAMS AND GLACIAL LAKES

Ice streams on the periphery release most of the ice and debris connected to the Antarctic Ice Sheet more than 90% of the ice sheet's total discharge, for instance, these ice streams are essential elements of the glacial system. Ice streams are small channels of swiftly moving ice that move at speeds between 0.5 and 1 km/year, which is more than twice as fast as adjacent ice (Weertman, 1972). As a result, ice streams usually have a lot of crevasses and sharp lateral borders, also known as shear margins, between them and the surrounding ice. Uncertain factors determine the position, dynamics, and behavior of these ice streams; but, in certain instances, they seem to cover sedimentary basins that could deform beneath the ice, promoting forward flow.

The knowledge that ice streams can turn on and off over time adds to this complexity. For instance, the Kamb Ice Stream in West Antarctica, formerly known as Ice Stream C, stopped flowing about 150 years ago and hasn't moved since. Ice shelves encircle about half of the Antarctic continent and are the floating extension of outlet glaciers.

A few of these ice shelves are massive; the Ross Ice Shelf, for example, covers an area roughly equal to that of France. Calving of icebergs from ice-shelf termini accounts for about 90 percent of Antarctic ice loss. Ice shelves play a crucial role in influencing the dynamics, and therefore the response time of upstream ice from inland Antarctica, because they delay ice flow from the glaciers or ice streams that feed them. The ice shelves contribute a significant amount of fresh water to the oceans because of the quick heat exchange. As calved icebergs drift to sea, ice shelves can direct, transport, and deposit substantial amounts of glacial sediment. This allows the sediment to be dispersed more widely.

When glaciers reach sea level, they end as tidewater glaciers and calve quickly to provide another effective way for the ice sheet to lose ice. In these cases, the glaciers do not extend to form an ice shelf. There are concerns that some glaciers, including Pine Island Glacier and the nearby Thwaites Glacier, are changing their velocity structures quickly and irreversibly (accelerating at rates of up to 5% faster/year). This is concerning because, should this trend continue, the ice sheet may become unstable due to the drawdown of ice from interior Antarctica, which might result in a sharp rise in sea levels.

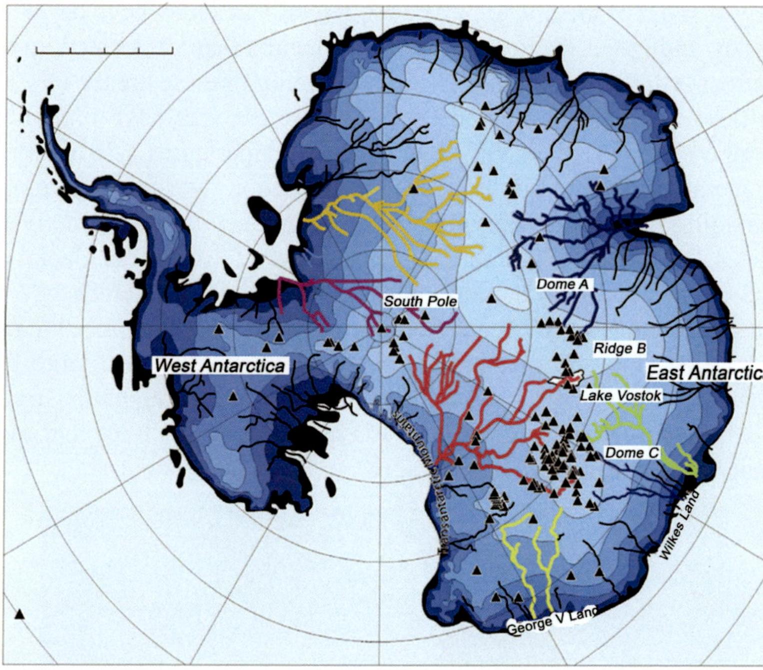

Figure 2.2 Subglacial lakes

Source: https://commons.wikimedia.org/wiki/File:Antarctic_subglacial_lakes.png, Creative Commons License.

Subglacial lakes (Figure 2.2) are huge freshwater reservoirs, situated beneath the ice sheet, and are located in regions where topographic depressions are overlaid by basal melting zones resulting from elevated geothermal heat flow. Currently, more than 70 subglacial lakes under the Antarctic Ice Sheet have been found. The largest is Lake Vostok, which is 230 km long, 14,000 km^2 in area, and has a water volume of about 2000 km^3; it is located beneath 3 km of ice (Siegert, 2000)satellite altimetry was used to measure anomalous near-flat regions on the ice-sheet surface that represent a manifestation of the subglacial lake beneath. Using RES and satellite altimetry, the location and extent of Antarctic subglacial lakes can be identified. The largest subglacial lake exists beneath Vostok Station, and is 14,000 km2 in area. The combined area of additional subglacial lakes beneath Dome C is 15,000 km2 and at least 15,000 km2 of lake surface lies beneath the remainder of the ice sheet. The water depth of subglacial lakes can be estimated through seismic investigations (although data exist only for Lake Vostok. The lakes are connected subglacially to one another. Large meltwater channel systems are evidenced by landforms that suggest some of the old Antarctic subglacial lakes may have drained in the past. Valley and outlet glaciers are smaller glaciers that are located away from the ice streams that flow quickly. The most familiar of these are the Dry Valleys, as they flow into regions devoid of ice. These glaciers have extremely low basal temperatures (-15°C), very low surface temperatures (-30°C), and little to no meltwater contained in the ice. Their accretion rates are also remarkably low, averaging only a few centimeters per year. They have extremely low ice velocities as a result, and as a result, the area has relatively less glacial erosion and land expansion. This habitat actually has a record of landscape change that dates back millions of years because rates of landscape modification are so low here. Cold-based glaciers (Figure 2.3) frequently have large loads of basal debris, particularly close to their margins where regelation frequently causes debris to pile and produces thick layers of debris-rich ice that can reach a thickness of five meters.

Figure 2.3 Cold-based glaciers

Source: https://commons.wikimedia.org/wiki/File:Glaciers_At_El_ Chalten_%28190265829%29.jpeg,Creative Commons License.

2.4 GREENLAND

The second-largest ice sheet in the world, the Greenland Ice Sheet is thought to be particularly susceptible to global warming. This is due to two factors: first, it covers a latitude range of 60 to 83 degrees North; and second, it is surrounded by ocean waters, which, in contrast to Antarctica, lack ocean currents that would isolate the ice sheet from heat transfer from mid-latitude regions. Research indicates that the ice sheet has undergone extraordinary levels of surface melting in the past few years. In the last ten years, there has also been an acceleration and thinning of several of the outlet glaciers that drain Greenland's Ice Sheet. The annual net volume loss of the Greenland Ice Sheet has increased from 90 to 220 km3. The acceleration of multiple large outlet glaciers is the primary driver of this surge.

Figure 2.4 Kangerlussuaq Glacier

Source: https://commons.wikimedia.org/wiki/File:Kangerlussuaq_Glacier_Aug15_2018.jpg, Creative Commons License.

Particularly, three outlet glaciers, Jakobshavn Isbrae, Helheim Glacier, and Kangerlussuaq Glacier (Figure 2.4) have experienced remarkable decline, surface lowering, and speeding up. These alterations are significant because the three glaciers collectively remove about 12% of the ice from the Greenland Ice Sheet (Young et al., 2020). Large amounts of ice may be drawdown by the glaciers from the interior of the ice sheet, and more ice flowing into the oceans would inevitably raise sea levels. The Zwally effect is the process, which depends on meltwater penetrating the ice sheet bed and raising basal water pressure to reduce friction. The mechanism is simple. When air temperatures rise, surface melting increases, increasing the amount of water on the surface of the ice sheets. This surface meltwater offers lubrication and lowers friction as it drains through crevasses, moulins, and other cracks to the glacier base, increasing the sliding velocity. According to the Zwally hypothesis, the relationship between surface melting and ice-sheet flow offers a mechanism for the ice sheets' large-scale, dynamic, and quick responses to global warming.

Terry Hughes invented the term "Jakobshavn effect" to characterize a scenario in which a minimal force imbalance in a calving glacier results in a significant reaction in the glacier behavior (Hughes et al., 2016)as a thawed bed is progressively drowned for stream flow, and as lateral and/or local grounding vanish for shelf flow. This can reduce ice thicknesses by 90 % and ice elevations by 99 % along flowlines. Original work presented here includes (1. The glacier may become more buoyant, as a result of thinning at the calving front, increasing its sensitivity to tidal fluctuations. The pace of iceberg calving increases as a result of reduced friction brought on by increased buoyancy, which accelerates the drawdown and acceleration of glaciers. The unusual thinning of the outlet glaciers has been attributed to this phenomenon. Additional factors that could cause disturbances to tidewater glacier behavior include shifting marine impacts, like rising sea temperatures and the melting of sea ice in fjords. The mountainous regions of New Zealand and South America are home to several ice caps, outlet glaciers, and isolated valley glaciers. The western glaciers are highly dynamic and have substantial mass-balance gradients. Eastern glaciers are less active due to strong west-to-east precipitation gradients. Within the Southern Alps, between 43- and 45 degrees South, is where most of New Zealand's glacialization occurs. For instance, the Franz Josef Glacier experiences annual precipitation of over 5000 mm, reaching a maximum of 15,000 mm in the neve above the equilibrium line (Schaefer et al., 2015)surface and deep ocean, hydrology, and cryosphere, which is fundamental for evaluating the robustness of earth's climate in view of ongoing climate change. The Last Glacial Maximum (LGM, 26–19 ka ago. As a result, the Franz Josef Glacier centerline velocities are measured to be 600 meters each year. The climate of the glaciers east of the Main Divide is significantly drier. Another area that is well-represented by temperate glaciers in Iceland, which has been the subject of glacial geology research for the past 30 years. Specifically, research has been done on the southern outlet glaciers of the Vatnajökull Ice Cap. The Himalayas stretches 2000 kilometers from Burma to Afghanistan. Although it is one of the world's most glacierized locations outside of the polar regions, less is known about it from the standpoint of glaciology. The monsoon and the mid-latitude westerlies are the two main weather systems that affect the Himalayas. The glaciers in the Everest region are likely the most researched because of the South Asian monsoon from the Indian Ocean, which has a strong influence there and causes a summertime maximum of precipitation that lasts from late May to mid-September. The range of elevations where maximum precipitation occurs is 5000–7000 m. Even though these peaks experience below-freezing daily temperatures for a large portion of the year, strong winds keep large amounts of snow from accumulating at high elevations. In contrast to polar or lower-altitude alpine glaciers, the peak of accumulation and ablation happens at the same time because of the coincidence of the highest temperatures and maximum precipitation. The glaciers frequently have a large supraglacial debris load made up of this rockfall material.

Figure 2.5 Moraines

Source: https://commons.wikimedia.org/wiki/File:Glacial_moraines_in_Alam_Chal_Hadi_Karimi.jpg, Creative Commons License.

2.5 MATERIALS

From the Ice Age or earlier, some of the glaciers have lateral and frontal moraines (Figure 2.5) that can reach heights of 100 meters (Evans, 2009)moraine linearity has a low preservation potential in deglaciated terrains but hummocky moraine tracts previously interpreted as evidence of areal stagnation may instead record receding polythermal glacier margins in which debris-rich ice was concentrated in frozen toe zones. Recent applications of modern glaciological analogues to palaeoglaciological reconstructions have implied that: (a. The moraines frequently have an unvegetated, loose, disintegrating interior and a vegetated exterior. In extreme conditions, terminal moraines can rise to 250 meters above the surrounding terrain, as is the case on Khumbu Glacier (Figure 2.6).

Figure 2.6 Khumbu Glacier

Source: https://commons.wikimedia.org/wiki/File:Khumbu_Glacier,_Everest_Zone,_Nepal,_Asia.jpg, Creative Commons License.

Sand-gravel from a combination of rockfall debris (Figure 2.7) and material from the valley sides makes up the majority of the material that makes up terminal and lateral moraines.

Figure 2.7 Rockfall debris

Source: https://commons.wikimedia.org/wiki/File:2012_Siachen_Glacier_avalanche_02.JPG,Creative Commons License.

The Himalayan glaciers have surface velocities of between 10 and 100 meters per year when they are active. However, the flow rates of many Himalayan glaciers are far slower, particularly in their terminal regions. As a result, many glaciers are experiencing down-wasting and becoming debris-mantled in their lower reaches rather than actively receding. The majority of the debris layer is made up of coarse, angular particles and is hummock. There are depressions with meltwater ponds between the debris hummocks, as well as steep exposed ice slopes covered in debris or ice cliffs several meters high. The surface gradient is lowered or even reversed as a result of the heavier debris cover near the terminus, which delays ablation more than the thinner debris layer up-glacier. Large supraglacial lakes are formed over time when these lakes combine and are dammed by lateral and terminal moraines. These lakes create a natural risk where they are dammed by ice-cored moraines because the dams can suddenly break and trigger Glacial Lake Outburst Floods (GLOFs). In their proximal zones, glacial rivers in the Himalayas frequently lack broad braided reaches.

Therefore, in these high-altitude habitats, glaciofluvial processes are rather insignificant when compared to alpine regions. Because of the narrow valleys and steep channel gradient, meltwater is mostly limited to single channels.

The footprints of previous GLOFs are represented by well-developed alluvial fans and ruptured terminal moraines beneath several of the glaciers. Peak GLOF discharges can be up to 60 times larger than seasonal high-flow floods caused by runoff from snowmelt, glacier meltwater, and monsoon precipitation. Because of this, these floods are strong and may carry, deposit, and erode vast amounts of silt across several kilometers.

Figure 2.8 Andes of South America

Source: https://commons.wikimedia.org/wiki/File:Andes_Mountains_South_America_Photograph_020.JPG,Creative Commons License.

2.6 TROPICAL GLACIERS

The Andes of South America (Figure 2.8), spanning from Bolivia to Venezuela, contain more than 99 percent of the tropical glacier area, with the Peruvian

Cordilleras alone accounting for more than 70 percent. Because the tropical climate differs greatly from that of the mid-latitude or polar regions, glaciers can only be found at these low latitudes in locations where there are high mountains to permit snow accumulation. These mountains have a uniform climate, devoid of activity, without thermal seasonality, and with clearly defined seasons of precipitation. For instance, the year can be split into three climatic periods around the tropical glaciers of the South American Andes: the dry season (May–August), the wet season (January–April), and a transition period (September–December) during which precipitation gradually increases towards the wet season. Mass balance relationships on these glaciers are complicated because, in many locations, the maximum precipitation and snow accumulation periods also correspond with the highest air temperatures and ablation rates. An excellent illustration of tropical glaciation can be seen in the Peruvian Cordillera Blanca glaciers. With about 720 km^2 of high-altitude glaciers, this region is the largest glaciated area in the tropics, accounting for 25% of all tropical glaciers (Owen & Benn, 2005)but were estimated using a variety of methods some of which are not appropriate for high-altitude glaciers which are strongly influenced by topography and have extensive supraglacial debris cover. Furthermore, few of the reconstructions have adequate chronological control because of the difficulty of dating glacial landforms in high-altitude regions where organic matter for radiocarbon dating is sparse or absent. Cosmogenic radionuclide surface exposure and optically-stimulated luminescence dating are providing data to define the timing of glaciation and these, together with objective analysis of glacier dynamics, may allow better estimates of ELA changes in the future. However, there are currently only two regions (the Hunza Valley and the Khumbu Himal.

Because of the extremely low air temperatures at high altitudes, glaciers can be found here. Tropical glaciers do not have summer melt seasons that are marked by diverse temperatures above freezing, in contrast to mid-latitude glaciers. While annual exposure to air above freezing directly warms their lower altitude regions, ablation at higher elevations is primarily sustained by sunlight absorption. The amount of solar radiation absorbed, air temperature, atmospheric humidity, cloud cover, and wind all affect ablation. Because they respond to variations in ablation more quickly than glaciers in mid-latitudes, tropical glaciers have experienced significant retreat. The nineteenth and twentieth centuries have seen a decline in the volume of the tropical glaciers in the Cordillera Blanca due to their rapid retreat in recent years. It has been demonstrated that larger-scale climatic phenomena, like the El Nino-Southern Oscillation (ENSO), affect the interannual variability of temperature, precipitation, and stream discharge in this area. A portion of the glacier shrinkage can be attributed to rising air temperatures. The diverse landforms and materials produced by tropical glaciers are remarkable. In the tropical Andes, the majority of glaciers are retreating and experiencing down-wasting. Similar to those formed on Himalayan glaciers, the debris creates an uneven blanket of hummocks, coarse angular material scattered with supraglacial ponds.

In fact, the cleavage of the Arctic ice shelf provides evidence that environmental change has been accelerated recently. The Ward Hunt Ice Shelf, which covers 443 km^2, was a large ice shelf that was located along the Ellesmere Island coast in Nunavut, Canada, until recently. Between 2000 and 2002, the ice shelf abruptly broke apart following a 30-year decrease in its extent (Van Wychen et al., 2020). The greatest ice shelf in the Arctic disappeared at this point.

Figure 2.9 Arctic glacier

Source: https://commons.wikimedia.org/wiki/File:Monacobreen_Glacier,_Svalbard,_Arctic_%2820284919445%29.jpg,Creative Commons License.

Many Arctic glaciers (Figure 2.9) are polythermal, meaning that while the thicker ice in the accumulation area is frequently warm, the glacier lateral borders and surface layer are below the pressure-melting threshold. The dynamics of polythermal glaciers are complicated by this mixture of thermal regimes. The normal polythermal glacier moves only via internal deformation where cold-based ice predominates at the glacier terminus and lateral margins, and either by sliding on its bed or by deformation where warm-based ice dominates in the accumulation zone. Supra and englacial pathways are typically followed by meltwater. Consequently, polythermal glaciers typically have significant basal debris loads, with up to 50% of their surface area composed of debris-rich basal ice zones that are between one and three meters thick. Although medial moraines are frequently seen in their lower ablation areas, their surfaces hardly ever have a layer of debris covering them. Svalbard is well known for having a large percentage of surge-type glaciers. These glaciers are prone to sudden spikes in velocity, interspersed with periods of latency. The surface of these glaciers frequently has layers of ice created by refreezing melted snow. This typically develops on the surface of glaciers during brief bursts of warm weather in the early winter, which frequently coincide with precipitation and accelerate glacier melting.

Figure 2.10 Runoff from glaciers

Source: https://commons.wikimedia.org/wiki/File:Glacial_Run-off_%28182482127%29.jpeg,Creative Commons License.

2.7 RUNOFF FROM GLACIERS

Runoff (Figure 2.10) from glaciers is a valuable resource for the global economy in many regions, since it provides hydroelectric power, drinking water, and irrigation for crops. Meltwater is the primary ablation product on many glaciers. For example, the tributaries of the glacier-fed Indus and Ganges rivers provide drinking and irrigation water to an estimated 500 million people (Molden et al., 2022)with increased warming and accelerated melting of snow and glaciersGlaciers. This threatens the water, food, energy and livelihood security for many in Asia. The links between mountains and plains and the differential impacts of climate changeClimatechange on societies upstream and downstream need to be better established to improve adaptation measures. This chapter sheds light on climate changeClimatechange impacts on the cryosphere and mountains, the impact on river systems and the social consequences of such changes in mountains, hills and plains. In high mountains and hills, the impact of climate changeClimatechange is clear, as seen in changes in agropastoral systems and the increasing occurrence of floodsFloods and droughtsDrought, with losses and damages already high. Moving downstream, the climate changeClimatechange signal is harder to separate from other environmental and management factors. This chapter outlines how climate changeClimatechange in the mountains will impact various sectors in the hills and plains, such as hydropower, irrigationIrrigation, cities, industries and the environment. It discusses how climate changeClimatechange will potentially lead to increased disasters and out-migrationMigration of people. The chapter concludes by highlighting necessary actions, such as the need to reduce emissions globally, build regional cooperation between HKHHindu Kush Himalayas (HKH. A key factor regulating the dynamics of ice sheets and glaciers is glacier hydrology. Because subglacial meltwater is necessary for subglacial sediment deformation and because glacier motion, such as accelerated glacier sliding, is closely linked to water flow through glaciers. A glacier hazard may arise from the unexpected discharge of water that has been held inside or next to a glacier. Lakes with ice or moraine dams that drain quickly can endanger downstream residents and infrastructure. Debris from the ice-rock interface is removed by glacial meltwater, which then transports it outside the glacier boundaries and deposits it there. The dispersion of these glacial meltwater sediments creates a substantial economic advantage in many glaciated regions of the world. Important water-bearing aquifers can also be found in sand and gravel layers. It is crucial to understand that meltwater can be stored in glaciers in a variety of ways in addition to flowing through them. Storage takes place on several distinct temporal and spatial scales as ice, snow, and water. Meltwater can be stored in subglacial lakes, subglacial pore spaces in sediments, and cavities in subglacial locations. It can also be kept in supraglacial lakes, on the surface of glaciers as snow and firn, and within the ice in englacial water pockets, tunnels, and cavities.

Long-term storage takes place as firn and glacier ice, lasting years or even centuries. This storage can change the global sea level and has an impact on the long-term water balance of glacierized catchments. Snow and water are seasonal materials that are stored and released for periods ranging from days to years. The runoff features of glacierized catchments and the downstream river flow regimes are impacted by intermediate-term storage. The daily effects of drainage through a glacier, such as meltwater routing via snow and firn, as well as englacial and subglacial channels, are included in short-term storage (hours to days). Together with these time-dependent processes, there are also event-driven storage releases, such as the drainage of subglacial, moraine or glacier-dammed water bodies and floods. From its point of origin, glacial meltwater travels through the glacier via many distinct flow paths.

Figure 2.11 Storglaciaren

Source: https://commons.wikimedia.org/wiki/File:Storglaci%C3%A4ren_weather_station.JPG, Creative Commons License.

Both surface and basal melting produce meltwater, and while they usually include channel flow either within or on the glacier, they tend to travel various routes through them. The kind of channel system found in a glacier is mostly determined by its temperature regime. Meltwater can go through warm ice glaciers without freezing, allowing water to flow through supraglacial, englacial, and subglacial channels. More intricate flow patterns are present in glaciers with varied thermal regimes, where water can pass through supraglacial, englacial, and subglacial channels.

2.8 THE ROLE OF DEBRIS

Extensive research has been conducted on the internal structure of moulins and englacial drainage channels on the debris-covered Himalayan glaciers and the Swedish glacier Storglaciaren (Figure 2.11). The position of englacial drainage channels on the Himalayan glaciers is dictated by pre-existing paths with strong hydraulic conductivity. Storglaciaren moulins are likewise regulated structurally. A supraglacial meltwater stream is frequently interconnected by a crevasse that opens on the surface of the glacier. Subsequently, water will fill the crevasse until it opens up and deepens enough to cross englacial drainage channels, at which point the water will drain out. The crevasse may close when glacier flow pushes it into a compression zone, but the heat from the melting water may keep the drainage channel open, creating a moulin in the process. The Storglaciaren moulins are made up of vertical shafts that are between 30 and 40 meters deep. These shafts provide englacial tunnels that are at angles from 0 to 45 degrees from the moulin base. The original crevasse from which the moulin originated frequently controls the orientation of the englacial tunnels. Glacial erosion refers to the processes by which a glacier separates and transports rock and sediment as it moves. Because they carry material within them as they flow towards their borders, glaciers and ice sheets are agents of erosion. Either subglacial tunnels or ice-margin voids provide evidence of glacial erosion processes. Glacial abrasion is the gradual erosion of the bed caused by the passage of ice-carrying debris, and glacial plucking or quarrying is the extraction of rock or sediment in the form of blocks from the glacier bed.

The process of rock fragments being carried across a bedrock surface by a glacier cuts the surface. This is known as glacial abrasion (Boulton, 1982). This procedure is frequently compared to the action of sandpaper being scraped across a wooden board. The bed is predominantly eroded by large clastic rocks (0.01 m) embedded in the glacier, which generate grooves. On the other hand, finer material, particularly the silt polishes the surfaces of the bedrock. Although field measurements are still needed to prove this, there is strong theoretical evidence that the erosion of bedrock surfaces is significantly impacted by the polishing of the surface by fine material than by the scratching of the surface by larger clastic rocks. Three key factors influence glacier capacity to erode its bed. These are: 1. the pace of basal sliding; 2. the basal contact pressure; and 3. the concentration and supply of rock fragments within the glacier base. Some opposing theories, however, exist regarding the mechanisms governing this contact pressure and basal clastic rock migration. Because cold-based glaciers have less energy to erode their beds and because sliding is not common beneath them, the basal thermal regime is significant. The rate of abrasion is also determined by the debris concentration in the basal ice. Debris within the ice is what allows it to abrade significantly; ice alone cannot do this. However, merely raising the concentration of basal debris does not raise the rate of abrasion. Indeed, it has been argued that abrasion works best in situations when there is comparatively minimal basal debris. This is because the presence of basal debris lowers the sliding velocity by increasing the friction drag between the ice and its bed. A glacier can move more quickly if its basal ice is relatively clean than if it has a lot of debris in it. Because debris has a negative impact on the rate of basal sliding, there is a specific concentration threshold at which the abrasion rate decreases with increasing debris content. Other significant aspects of basal debris are its form and kind. Because erosion would diminish the size of the basal rocks first, minimal abrasion would occur if the debris was softer than the substrate. Because sharp fragments can pierce deeper into the underlying bedrock than those with rounded tips or edges, the shape of the basal debris is also significant. Basal debris needs to be continuously supplied for abrasion to be effective. Rock surfaces may expand and fracture when they become unloaded.

2.9 GLACIER MOVEMENT

A glacier movement across an uneven bedrock surface creates intricate patterns of basal ice pressure (Bennett, 2003)like the arteries in our body, their behaviour and stability is essential to the well being of an ice sheet. Ice streams may either be constrained by topography (topographic ice streams. Stress forms inside the bedrock as a result of this pattern of pressure differentials to the bedrock underneath. The bedrock may fracture as a result of these stresses. A discontinuous rock mass is one that has these kinds of vulnerabilities. Every piece of bedrock has joints, bedding planes, and other weak points that could be extended by the stress. The fact that the pattern of joints, and bedding planes within the parent rock mass controls the morphology of numerous glacial erosional landforms reflects the significance of these faults. Variations in the temperature of the basal ice can also be caused by temporal variations in the ice pressure pattern, and in certain situations, these variations can result in the formation of small cold areas within a warm-based glacier. Ice melting releases latent heat and uses thermal energy, which cools the ice mass. A portion of the meltwater will flow beneath the glacier to regions with lower basal ice pressure, where it will refreeze. Refreezing of the meltwater will occur if the basal ice pressure drops. Latent heat released during freezing will cause the basal ice to warm. Nevertheless, the temperature of the basal ice cannot return to its previous value due to the loss of some meltwater, and a cold patch will develop. To get the basal ice back to its original temperature, the same amount of water would need to refreeze as was melted, but since the meltwater has already been released, this cannot happen. As a result, there may be cold spots on the glacier bed due to temporal pressure variations beneath an ice sheet, which could be caused by changes in the ice sheet velocity or the release of meltwater. Variations in basal water pressure may also aid in the propagation of bedrock fractures beneath a glacier. Basal water pressure plays a key role in identifying whether or not basal cavities exist beneath a glacier.

The hydrostatic pressure created when a hollow forms as a result of an increase in water pressure can effectively raise the glacier base. As the glacier base rises, any rock fragments that are frozen to the glacier bed as a result of the heat-pump action during this process will be lifted and transported onward. Erosion cycles can occur in situations where variations in basal water pressure are coupled with the heat-pump effect. A phase of fracture propagation may be brought on by an increase in effective normal pressure over an obstruction when basal water pressure lowers. The equilibrium between the frictional forces that work to hold the rock in place and the tractive force that the overlying snow applies to it controls the evacuation and entrainment of a rock fragment from the glacier bed. It could be possible to reduce the frictional forces holding debris in place by having high basal water pressure. By drilling or tunneling to the base of a glacier, one can perform direct observations. Measurements of glacial abrasion in these tunnels can be made by periodically observing the microtopography of bedrocks. Glacial abrasion rates under Iceland glaciers have been shown to range from 0.5 to 1 mm per year based on observations of glacial erosion (Hallet et al., 1996). Glacial abrasion tends to concentrate in certain regions when basal debris is arranged into discrete lines or streams at the base of a glacier, allowing the snow to create channels or grooves. The other mechanisms consist of either a hyperconcentrated flow of water and meltwater processes. The size of the erosional mark created is determined by the obstacle size. They can develop around individual crystals, grains, or nodules that peek through the surface of the bedrock, or they can form around larger bedrocks. Unlike other landforms primarily formed by glacial abrasion, stoss and lee have a distinct asymmetry. They are described as bedrock knolls with a steeper, coarser slope that is plucked or quarried on the gently abraded slope. A roche moutonnée is the most prevalent kind of stoss and lee landform. These landforms can range in size from a few meters to hundreds of meters, and they are frequently seen. The presence of subglacial meltwater aids in the construction of glacial quarries, just as regular and frequent variations in basal water pressure help glacial quarrying.

2.10 MORPHOLOGY OF THE ROCHE MOUTONNÉE

The specific morphology of the roche moutonnée is determined by the characteristics of the parent bedrock once a lee-side hollow has opened and glacial plucking has begun. Because depth and spacing dictate the size of the blocks that can be quarried from the lee of the original bedrock hummock, bedrock jointing is very significant. The evolution of the plucked surface of a roche moutonnée can be predicted. Because of variations in basal pressure, a glacier may cause its own fractures in lightly joined bedrock. In this instance, there may be greater variation created by removing blocks. The surfaces of Rocheuses Montagnes are frequently home to a variety of microscale erosional landforms. Although quarried lee sides and abraded stoss slopes are characteristics of roche moutonnée, the orientation of these surfaces is not always a reliable indicator of the direction of snow movement because the pattern of joints or other weaknesses within the rock mass can affect their detailed morphology. Their accuracy as markers of palaeo-ice flow is so diminished. Either glacier abrasion or meltwater erosion can create grooves. The morphology of grooves and striations is similar, with the exception that grooves are deeper and larger. They can be up to several meters wide and one meter deep, with lengths ranging from tens to hundreds of meters. They are most likely the result of glacial erosion, while meltwater flow may also have an important role. The existence of structural flaws in the rock mass may also have an impact on the orientation and placement of specific grooves. Rock basins are discrete indentations etched into the bedrock. They can fill with water during deglaciation to form lakes, and they are frequently seen close to roches moutonnée. The diameter of a rock basin can range from 10 meters to hundreds of meters. The distribution of structural defects in a rock mass that can be used by glacial quarrying governs how these basins emerge. Therefore, the distance between joints within the rock mass usually determines the basin size and density.

Where ridges of englacial debris are exposed by surface melting, ablation-dominant medial moraines occur. In the ablation zone, these moraines seem to grow out of the glacier surface. Debris may follow a straight path down the glacier from a point where it is supplied by a nunatak onto the glacier. When the accumulation zone fills with snow, the debris will gradually move through the glacier in the direction of flow. Melting in the ablation zone will expose the debris as a medial moraine on the surface of the glacier. Debris-rich snow can fold via foliation and stratification to form medial moraines(Hart, 1998). Debris may accumulate as a diffuse layer across a large portion of the glacier surface in the accumulation zone, where it will eventually be covered by snow and create debris strata. During glacier movement, debris layers may fold across the glacier, especially if the snow flows from a large basin into a narrow stem and is pushed laterally. The anticlinal crests of the longitudinal debris-rich folds are visible when surface melting lowers the glacier surface when the debris-rich snow reaches the ablation zone. The degree of transverse compression within the glacier controls the folding intensity. Along the axis of each fold, a number of small medial moraines may appear if the folding is open. On the other hand, if the folding is perfect, the debris within each fold can combine to produce a single medial ridge. The rate of debris supply in relation to the glacier flow velocity has an impact on the structure of debris on its surface. A thick coating of debris may build up if the glacier pace is low and the rate of debris supply is significant. The debris cover will spread across the glacier surface more quickly and thinner if the rate of debris supply is low and the glacier velocity is high. Because supraglacial debris acts as insulation and slows down surface melting, the presence of debris on a glacier's surface has a substantial impact on the mass balance of the glacier.

CHAPTER 3

GLACIAL TECTONICS

Contents

3.1 INTRODUCTION

Glaciotectonism is a crucial part of contemporary glacial theory. In southern Scandinavia and western Europe in the late 18th and early 19th centuries, glacial theory started to take shape (Nesje et al., 2008)records of Holocene glacier variations in different regions in Norway have been synthesised. During the period from approximately 8000 to 4000 cal. yr BP, most glaciers in Norway were completely melted away at least once due to high summer temperatures and/or reduced winter precipitation. Lichenometrically and historically dated moraines at Jostedalsbreen, in Jotunheimen, at Hardangerjøkulen, and at Folgefonna were used to extend records of glacier length variations back to their maximum position during the 'Little Ice Age'. The timing of the maximum 'Little Ice Age' glacial advance in different parts of southern Norway varied considerably, ranging from the early 18th century to the late 19th century. Cumulative glacier length variations of glaciers in southern Norway show an overall retreat from ~AD 1750 to the 1930s–40s. Thereafter, most Norwegian glaciers retreated significantly. Short maritime outlet glaciers with a short response time (<10–15 yr. Beginning in the 1760s, Horace-Benedict de Saussure was one of the first naturalists to conduct systematic observations of glaciers in the Mont Blanc region (Figure 3.1) (Carozzi, 1984).

Figure 3.1 Mont Blanc

Source: https://commons.wikimedia.org/wiki/File:Mont_Blanc,_Saint-Gervais-les-Bains_%28P1080018-HDR%29.jpg,Creative Commons License.

He saw different kinds of glaciers, as well as their movement and the source of moraines. The phrases "Roches moutonnée" and "moraine" were first used in geology by Saussure. The granite stones in the Jura Mountains were carried by glaciers from the Alps, as James Hutton reported in 1795 (Challinor, 1954). Jens Esmark concluded that glaciers had originally covered a substantial portion of Norway and the sea floor and that they had once been much larger and denser. He linked glacial deposition and transportation to erratic and moraines. Esmark also realized that the Norwegian fjords had been carved by the strong erosion caused by glaciers. Jean de Charpentier compiled the scientific data supporting earlier alpine glaciation in the early 1830s. He noted the Swiss Alp moraines, striations, and erratics in addition to the glaciers that were already there. Louis Agassiz postulated in 1837 that the northern hemisphere was formerly covered in vast sheets of ice. In the 1840s, Agassiz conducted extensive research on the flow of glaciers on the Unteraar Glacier in Switzerland (Seligman, 1949).

Unlike the more widely accepted dilatation or regelation theories of the time, Forbes demonstrated that glaciers move in part due to internal deformation. Large, erratics, moraines, and abrasion marks on boulders and bedrock were recognized, and these features were interpreted using data on glacier movements. These characteristics could only be attributed to glaciation that was widespread. Thus, structures produced by glacial deposition and erosion, as well as other features resulting from glacial erosion, have always been the foundation of glacial theory. It took several decades for the researchers to know that glaciers could reshape shallow crustal rocks and sediments. Ice-pushed structures were identified in several locations during the late 19th century, including Skene in southernmost Sweden, Mons Klint in Denmark, and Ratigen in Germany, as well as the southern New England islands in the United States. Before this, other explanations for the evident structural disruptions at these locations had been proposed, such as landslides, disturbances by volcanoes, mountain formation, etc. In the late 19th and early 20th centuries, ice-shoved features from inner continental areas were recognized. Gilbert in 1899 saw bedrock formations on the southern shore of Lake Ontario that had been distorted by glaciers. A variety of ice-pushed constructions were reported in Poland and close to Minneapolis, Minnesota. George Slater was the first to specialize in the study of structures affected by glaciers (Lee & Phillips, 2013)focuses on the styles of deformation and tectonic imprint (folds, fractures, fabrics, foliations and lineations. He decided to write his doctoral dissertation at the University of London on the topic of ice-push deformation. In England, Denmark, Canada, the United States, and the Isle of Man, he conducted research on ice-shoved mountains. The name glacial tectonics was first used by him, and it is now commonly abbreviated as glaciotectonics. The mechanisms of glaciotectonic deformation are referred to as glaciotectonism. However, there is a great deal of ambiguity regarding the precise definition and application of the term glaciotectonic. Glacial tectonics was a term used by Slater to describe structural disturbances in bedrock as well as deformations in glacier ice, even though he did not define it.

3.2 TECTONICS

Two basic criteria are used to identify glaciotectonic structures and landforms: the existence of distinct, pre-existing bedrock or sediment bodies; and the existence of deformations brought forth by glaciers inside those bodies. Deformed structures can be created mostly in situ (autochthonous) or during the transit and deposition of a detached material (allochthonous). Examples of these structures include folds, faults, breccia, mylonite, and other disturbances. These deformations cannot be the product of pre-existing structures carried over from the parent material; rather, they must be the consequence of forces exerted by glaciers. In certain instances, glaciotectonism might be understood as a form of partial or incomplete glacial erosion of bedrock. On the other hand, in certain situations, it may also be seen as a type of glacial deposit. Nevertheless, neither method is totally adequate; glaciotectonic characteristics are those conditions in which glacier ice can alter pre-existing strata without erasing or fully destroying the rock or sediment. The physical characteristics of the substratum material largely determine whether glaciotectonic structures can form. In the correct types of bedrock or glacial deposits, the same dynamic conditions that elsewhere induce conventional glacial erosion and deposition also produce glaciotectonic characteristics. There is now a difference between sedimentology and glaciotectonic terminology that comes from structural geology.

3.3 SEDIMENTS

Sedimentary layers that are moderately consolidated comprise the majority of deformed material (Braathen et al., 2013)West Norway. These include gravel, sand, silt and clay, sandstone, chalk, lignite and coal, etc., and are usually Cenozoic or Mesozoic. Rarely have crystalline rocks and consolidated sedimentary strata experienced glaciotectonic deformation. These rocks, which include lithified limestone and dolostone, sandstone, gneiss and schist, granite, and other plutonic rocks, are often older i.e., Paleozoic, Proterozoic, and Archean. However, it should be acknowledged that the conditions at the time of deformation/transformation might have been very different from those that exist now. For instance, when the rocks were under strong pressure, several deformations occurred. Because there is so much variation in type and size, it is probably impossible to list every glaciotectonic structure. Depending on the type of deformation, glaciotectonic structures can be classified as brittle or ductile. Internal fluid flow of material causes ductile deformation. Since the rock mass has almost no internal strength during ductile deformation, even minute pressure fluctuations can alter its size or shape. Unconsolidated or fine-grained strata, such as shale, chalk, silt, or clay, that underwent deformation due to intense confining pressures are the most common types of ductile formations. Ductile deformation is represented by a variety of folds. When different strata of rock fracture and fail, brittle deformation occurs. The interior fabric of the rock between fractures does not exhibit permanent deformation; instead, deformation is achieved via movement or adjustment along fracture planes. The most common characteristics of brittle structures are seen in cemented or coarse-grained strata that underwent deformation from low confining pressure, such as sand, gravel, sandstone, limestone, or quartzite. Brittle deformation is seen in joints, faults, breccia, and fissures.

Underground fissures in the limestone were generated by glacial carving of well-consolidated Ordovician limestone layers along shale seams. These processes are based on a number of factors, including groundwater pressure, permafrost presence or absence, layer thickness, and lithology. Differentiating in thickness and lithology, layers within a stratified sequence react differently to deforming pressure. Consequently, certain layers exhibit ductile deformation while others produce brittle structures. Additionally, changes in pressure, temperature, or fluid content during deformation can cause a single rock to exhibit ductile structures in some areas while displaying brittle deformation in others. A few geologists have made an effort to categorize glaciotectonic formations based on factors such as genesis, size, style, and degree of deformation. Moran in 1971 distinguished between categories of glaciotectonic formations (Hart & Boulton, 1991)which results in glaciotectonic deformation, and also discusses the interaction of deformation and deposition within the subglacial environment. Two types of glaciotectonic deformation are discussed that are produced by the active movement of ice: (1. The surface or morphologic manifestations of subsurface structures brought about by glacial deformation of bedrock and glaciogenic strata are known as glaciotectonic landforms. The landforms might exhibit the primary morphology that they had when they were first formed during the glacial period. This is especially true for young landforms that have not been significantly altered by post-glacial erosion and deposition. In certain conditions, underlying features may be directly expressed in the morphology. In some cases, subsequent glacial or non-glacial events profoundly changed the glaciotectonic landforms. The current landforms might be mostly covered by more recent deposits, or they might be little more than the erosion-related remnants of the former shapes. The morphology in these regions only serves as a representation of the underlying components. In any situation, understanding stratigraphy and underlying structure is crucial for correctly interpreting the landforms.

Ice-shoved hills are made up of individual dislocated masses of drift and bedrock, referred to as floes, rafts, and megablocks. Both raft and mega blocks describe massive, relatively thin materials arranged essentially horizontally. Much like in other areas of earth science, the geometric analysis of glaciotectonic structures and landforms is done using similar methodologies. Most of the time, these techniques were created for different uses in the mapping and study of surface or subsurface geology before being used in glaciotectonic studies. Folds, faults, and distortions produced by other forms of tectonic deformation are essentially the same as glaciotectonic features. Indeed, glaciotectonic formations might be thought of as copies of actual mountains. Therefore, the study of glaciotectonic characteristics can be readily approached using the methods employed by structural geologists. The size, form, depth or thickness, and orientation of a structure must first be ascertained. The easiest way to accomplish this is in the field, where a range of approaches can be used, from basic surface exposure surveying to well logging and subsurface geophysical measurements. For additional study, the generated geometric data are then graphically shown as maps, cross-sections, block diagrams, or stereographic projections. The foundation of terrain analysis for structural purposes is the idea that topography, drainage, vegetation, and soils in a landscape frequently accurately represent the characteristics of subsurface structures. Topographic mapping is frequently used to map folds, faults, and other structures, even in cases when the bedrock itself is not readily apparent at the surface. The interpretation of vegetation and its relationship to geologic conditions is the foundation of the so-called geobotanic approach. Numerous geomorphic factors, including drainage, aspect and slope of the land, and sediment and soil, are integrated by vegetation and ultimately represent the underlying lithology and structure of the substratum. As a result, geomorphic conditions are frequently reflected in both natural and artificial vegetation patterns (Marston, 2010). This method works especially well for deciphering remotely sensed images, such as satellites.

A review of current, extensive topographic, geologic, and soil maps, along with aerial photographs of the area under investigation, is frequently the first step in the search for glaciotectonic events. A review could highlight landforms or geological elements that point to glaciotectonic characteristics. Such a review might or might not point to areas that warrant more research. In the field, one looks for accessible outcrops, excavations, and exposures, such as building sites, railroads, lakes and cliffs, river banks, sand and gravel pits, coal and clay mines, and so on. While some naturally occurring cliff parts may be accessible for a few days or weeks during a construction project, others may be available permanently. It is always beneficial to speak with nearby landowners, farmers, and ranchers as well as civil engineers, drillers, and other construction professionals. The majority are quite helpful and might allow access to study private property. A single, large, deeply exposed part is significantly more valuable than many small exposed sections. An in-depth study of all lithologic units and scrutiny of structures at all scales and levels within the geologic environment is made possible by the large fragment. If there isn't a chance to expose much ground, a section may be dug at a particular location that has been selected to maximize the possibility of uncovering subterranean strata and structures. In this instance, the excavation's angle, depth, and other features might be planned to provide the best chance for studying the buried structures and silt. Excavations can frequently be negotiated in conjunction with existing economic ventures. For instance, extra sections may be dug at building sites, coal mines, clay pits, etc. for a nominal fee. Similarly, water-well drillers are frequently open to going deeper to sample lower levels or reach bedrock hidden beneath sedimentation.

3.4 GEOLOGY

Most geologic field work requires a spatial precision of 1–10 meters, which can be achieved with inexpensive surveys. Advanced methods may be used, usually requiring more time and equipment for higher spatial accuracy (<1 m). Making geologic maps, gathering information on the orientation of formations, and measuring and sampling sections are some of the essential field operations. According to Davis and Reynolds, a stereographic projection is an effective tool for displaying and modifying the three-dimensional geometry of lines and planes. The line or plane's three-dimensional orientation is determined by the intersection it makes with the circumference of the sphere. The projection plane is a horizontal plane that splits the projection sphere in half. The projection plane indicates the azimuth directions. On the horizontal projection plane, your eye perceives an arc formed by the intersection of the dipping plane and the lower hemisphere. As a result, the resulting projection-plane arc is essentially a top-down perspective of the dipping plane. A collection of circular arcs that create parallels running E-W and meridians running N-S make up the Wulff stereonet. The stereonet is not a map, although the meridians and parallels resemble the longitude and latitude lines on one. The Wulff stereonet maintains angles; meridians and parallels always cross at right angles. Shape and size, however, are warped. Geologists have been interested in subsurface exploration for a long time, both theoretically and practically. As a result, a lot of work has gone into creating techniques for sampling and probing the interior of the Earth. These techniques can be classified into geophysical prospecting and direct sampling. Direct sampling involves drilling and recovery. Different geophysical methods can be used from the surface or inside of drill holes.

Ground penetrating radar or GPR (Figure 3.2), a geophysical technique, is one of the more recent technologies for subsurface prospecting (Loke et al., 2013). Short bursts of intense microwave radiation are sent into the ground from the surface in the case of GPR. The microwaves travel through the earth, engage in intricate interactions with subterranean materials, and can be reflected or scattered by surfaces that have different dielectric constants. Microwave frequency controls the spatial resolution and penetration depth. Deeper penetration is achieved at longer wavelengths (25 to 200 MHz), but resolution is decreased. To identify larger and deeper geologic features, GPR is helpful. Higher frequency, shorter wavelengths (300–1000 MHz) are more effective in detecting small, shallow objects, like individual rocks or pipelines. The water content of soil and sediment greatly affects microwave interactions; dry or frozen materials provide the ideal subsurface conditions. Seismic surveys could be useful tools for locating glaciotectonic features that are underground. Reflection seismology was originally developed for shallow geophysical prospecting (oil and gas). Large areas may be covered by ship-borne surveys in grid networks, which also offer high-resolution, multi-channel seismic data that can be used to analyze the stratigraphy and structures of the continental shelf. In the North Sea basin, Huuse and Lykke-Andersen mapped many widely distributed glaciotectonic structures in the mid-Miocene period (Huuse et al., 2001)near top Oligocene, the mid-Miocene unconformity, and base Quaternary. The maps show that the eastern Danish North Sea has been affected by faulting and salt diapirism throughout the Cenozoic. Carbonate mounds, erosional valleys and pockmark- or karst-like structures were identified at the top of the Upper Cretaceous–Danian Chalk Group. Strike-parallel erosional features and depositional geometries observed at near top Oligocene and at the mid-Miocene unconformity indicate that these major sequence boundaries can be attributed to large-scale lateral changes in sediment supply directions. Increases in sediment flux to the southeastern North Sea at the Eocene/Oligocene transition and in the post-Middle Miocene appear to correlate with similar events world wide and with long term δ18O increases, indicating forcing by global factors, i.e. eustasy and climate. Stratal geometries observed on the seismic data indicate that the so-called 'Neogene uplift' of the eastern Danish North Sea may have been hundreds of metres less than previously suggested. It is argued that late Cenozoic uplift of the basin margin and of mountain peaks in southern Norway may have been caused entirely by isostatic uplift of the crust in response to accelerated late Cenozoic denudation and dissection of topography created in the Paleogene. The late Cenozoic periods of accelerated denudation and incision rates were most likely driven by climatic deterioration and long term eustatic lowering rather than active late Cenozoic tectonics, the cause of which is conjectural. A series of shallow thrust structures and an associated system of deep, buried valleys were mapped. Thrust faulting most likely occurred in response to gravitational loading at the margin of an advancing ice sheet, and it was followed by deep incision due to subglacial melt-water erosion, probably during the Elsterian glaciation.","container-title":"Marine Geology","DOI":"10.1016/S0025-3227(01. When compared to findings from traditional, land-based geological studies, glaciotectonic formations found by offshore seismic data are usually larger in terms of area and depth, as well as having lower dip angles.

Figure 3.2 Ground penetrating radar

Source: https://commons.wikimedia.org/wiki/File:Using_ground-penetrating_radar_equipment_%2815894930731%29.jpg, Creative Commons License.

To precisely interpret glaciotectonic and stratigraphic conditions in three dimensions, data from boreholes, seismic and gravity surveys, and geoelectric profiles were combined with other geophysical surveying methods. Seismic profiles clearly captured complex glaciotectonic formations, their substrate, and load under challenging topographic settings. Dislocations appear to affect both Miocene and Oligocene formations, which were pushed primarily northeast, and they extend down to 300 meters. Traditional aerial photographs are large-format, vertical, panchromatic, and medium-scale. Aerial photos of glaciotectonic landforms have been used in geology atlases. For regional mapping and assessment of various environmental factors, including soils, geology, water resources, vegetation, etc., these aerial photos are essential. In theory, aerial photos of this kind are accessible for the majority of northern countries with noticeable glaciotectonic landforms; nevertheless, the cost, age, and quality of these images differ significantly.

Figure 3.3 Landsat

Source: https://en.wikipedia.org/wiki/Landsat_9#/media/File:LANDSAT-9.jpg,Creative Commons License.

3.5 SATELLITES

The Landsat satellites in the early 1970s marked the beginning of space-based remote sensing for civilian, and scientific purposes. A multitude of moderate-resolution datasets are available from Landsat instruments for a wide range of earth science applications. Many Landsat datasets are freely accessible and can be obtained for a nominal fee from other remote sensing sites like the EROS Data Center of the U.S. Geological Survey. Morrison and Slaney were the first to discover the potential of Landsat (Figure 3.3) imagery for glacial geomorphology. Due to low sun angle, snow cover, and a lack of active vegetation, winter Landsat images in low relief areas have a strong morphological character.

The RADARSAT (Figure 3.4) (Canadian Centre for Remote Sensing or CCRS) and the commercial high-resolution multispectral scanner IKONOS are two remarkable satellite systems. The latter's spatial resolution is comparable to that of conventional aerial photos. These days, a wide range of governmental and commercial sources offer a large number of additional systems and varieties of remotely sensed images obtained from space. Comparing remote sensing from space to traditional aerial photography reveals several benefits. The use of space-based remote sensing has transformed the field of geomorphology and has spawned a brand-new sub-discipline called megageomorphology, which is focused on the description and analysis of large-scale geomorphic features. This method has been applied to glacial geomorphology, where data from remote sensing has been used to map and interpret glaciated landscapes at scales ranging from large regions to continents. Punkari was one of the first to apply this method to the interpretation of glacial dynamics in Finland. In North America, comparable studies were conducted. Small format aerial photography (SFAP), on the other hand, has become a viable method for obtaining extensive, high-resolution pictures for in-depth site assessments (Fraser et al., 1995). This technique is based on compact digital or video cameras and 35- or 70-mm film cameras that are controlled from low-height, manned or unmanned platforms. There are several kinds of platforms that can be used, including kites, aircraft, and drones.

Figure 3.4 RADARSAT

Source: https://en.wikipedia.org/wiki/Radarsat-1#/media/File:Radarsat-1.gif,Creative Commons License.

Accurately mapping intricate glaciotectonic structures has been demonstrated to be possible with SFAP. Low-altitude and oblique air photographs can be used.

Photographs were taken from a small aircraft flying 200–300 meters parallel to the mountain. Standard vertical aerial photos of the research site are used to determine the ground control locations. Multi-model photogrammetric analysis can be performed on oblique stereopairs. Using this method, precise geological sections were built for the most famous glaciotectonic sites in Denmark. A 6 km long section with a spatial resolution of about 25 cm was created by Pedersen at a scale of 1:500 (Petersen et al., 2009). Such type of detailed geometric data is vital for explaining the glaciotectonic genesis of these important sites.

Because of the load, deeply buried rocks are subject to tremendous pressure that is constant and equal in all directions. This type of stress is known as lithostatic pressure, and the thickness and density of the load are the only factors that affect it. Glaciostatic pressure is the similar, constant force that glaciers and ice sheets apply to the substratum. Unlike tension, which is a negative stress, it is compressive or positive stress. Lithostatic pressure probably cannot explain bending and faulting, but it may cause strain in rocks as a result of simple compaction. Differential stress is necessary for this. Shear stress or uneven normal stresses can also result in the development of differential stress. Certain fundamental properties of material strength are revealed through laboratory testing of rock and sediment samples. Lithostatic compression applied to geologic materials results in a reasonably strong material. When differential stress is applied, shear stress is created within the rock, making it significantly weaker. Under strain, rocks and sediments become even more fragile. Shear stress or tension is what causes rocks and sediments to collapse structurally rather than compression.

3.6 LANDFORMS

Among glaciotectonic landforms, the hill-hole pair is perhaps the simplest. The concept of the hill-hole is repeated in other categories of glaciotectonic landforms. Jessen initially reported this relationship between distinct depressions and unique ice-shoved hills in northern Jutland, Denmark. Some researchers observed the pairing of anomalous hills with upglacier depressions in northern Dakota and accurately identified the glaciotectonic genesis of the hill-hole pairs. Depending on their internal composition, these hills were previously frequently mistakenly classified as kames. The shifting hills could be caused by bedrock or pre-existing drift. The material source that is now in the hill is represented by the depression. Nowadays, depressions are frequently the locations of lakes, bogs, bays, estuaries, or just low points in the surrounding terrain. Ideally, the volume of the depression and the volume of the slope should be about the same or slightly less. Determining a depression's initial volume, however, can be challenging because new sediment frequently fills depressions to some extent.

Sometimes it is impossible to locate a source of depression. The depression may exist, but it is concealed by recent sedimentation or is beneath a large lake or body of water. In a few instances, it may be shown that the displaced mass was actually an existing hill that was just shifted to a different spot. There are hills in Belarus and northeastern Poland without accompanying depressions. However, anomalous depressions without accompanying hills are also known to exist in several locations, including eastern Alberta and central Poland. The Devil's Lake Mountain is another example of a small hill-hole pair (Bickerdike et al., 2018). On its northwest side, the ice-shoved hill is a single, continuous ridge that is somewhat arcuate and runs parallel to a small source basin. The ice shoved ridge stretches 4 km in length and 1 km in width, with an elevation of up to 55 m above the source basin. A linear escarpment connects the ridges and the source basin edges, bounded by the hill-hole pair. The interpretation of this lineament as a fault validates the geometric relationship between the hill and the source basin. Even with a limited understanding of subsurface conditions, a hill-hole pair can be confidently identified when all morphologic characteristics are present. Ice-shoved hills are frequently smaller due to subsequent erosion or may be partially hidden by the accumulation of glacial silt.

Figure 3.5 Herschel Island

Source: https://commons.wikimedia.org/wiki/File:Permafrost_in_Herschel_Island_019.jpg, Creative Commons License.

Herschel Island (Figure 3.5) is situated in northern Yukon, Canada, at the western boundary of Mackenzie Bay in the Beaufort Sea. It is a section of the Yukon Coastal Plain, which is primarily <60 m in height and generally low and practically flat. The most famous is Herschel Island, which spans over 100 km^2 and has a maximum elevation of 181 m(Möller et al., 2010)Johannes V. Jensen Land, located on the north coast of Greenland. This is a land of extreme climate: polar desert conditions and a coast bound by a permanent sea ice cover. Our data covers the period from the Last Glacial Maximum (LGM, 18–22 cal ka BP. On its northern side, the island is surrounded by rugged sea cliffs that rise to a height of over 60 m. It is made up almost exclusively of Pleistocene sediments that were pushed by ice from the southeast and originated in the Herschel Basin. One of the massive hill-holes pairs in the world is made up of Herschel Island and Herschel Basin. A long ridge of comparable ice-shoved Pleistocene deposits is found with maximum elevations exceeding 75 meters.

Figure 3.6 Yukon Coastal Plain

Source: https://commons.wikimedia.org/wiki/File:Yukon_Delta,_Alaska_-_NASA_Earth_Observatory.jpg,Creative Commons License.

The Yukon Coastal Plain (Figure 3.6) or YCP is covered in tundra and has a continuous layer of permafrost that is often more than 300 meters thick underground. This permafrost consolidates the Pleistocene strata of the Kay-King Points Ridge and Herschel Island. The preglacial Pleistocene sediments of Kay-King Points Ridge and Herschel Island constitute the main component of the ice-pushed structures of these areas. In none of these locations is there any exposed pre-Quaternary bedrock. Before the Buckland Glaciation, shallow marine, lacustrine, delta, lagoon, or flood-plain settings produced the interbedded clay, silt, sand, gravel, and peat that make up the deformed layers. Glaciotectonic deformations, and intermittent exposures hinder the regional connection of these strata.

Since the early 1800s, Devils Lake, a large natural lake in northern North Dakota, has been the subject of intense scientific investigation (Todhunter, 2021). Glacier pushing significantly altered the northeastern North Dakota topography, with Devils Lake serving as a notable example. The lake is located in a network of interconnected depressions created by glaciers pushing gravel and shale from the Cretaceous bedrock. During the Wisconsin glaciation, debris from the Devils Lake depressions formed Sullys Hill, and other hills to the south. A breccia mixture of deformed shale and glacial silt makes up the ice-shoved hills. There are more than 200 meters of relief between Sullys Hill and the bottom of central Devils Lake. One of the largest and well-developed glacial landforms of the hill-hole type in the United States are the ice-scooped depressions and ice-shoved hills. Under the Devils Lake area, an important aquifer known as the Spiritwood aquifer trends from east-southeast to west-northwest. This aquifer was created in the preglacial Cannonball Valley, partially in bedrock and partially in loose sediment. This aquifer is situated above Devils Lake's primary depressions. Major glaciotectonic thrusting was probably easier because of the high ground-water pressure in the Spiritwood aquifer. It is possible to interpret these trends of ice-shoved hills as structures created by thrusting at the edges of ice lobes that are converging. Sullys Hill is the merging point of two lobes, one of which traveled from the northeast in the East Bay/Devils Heart Butte region and the other from the north or northwest in the West Bay/Crow Hill area. The two ice lobes most likely moved toward Cooperstown simultaneously. The primary means of interpreting glacial geomorphology include visual inspection of Landsat images, color-infrared aerial photos, traditional topographic maps, and ground observations.

Sullys Hill serves as a central point for the surrounding glacier features. At Sullys Hill, two groups of ice-shoved ridges and related source depressions, or lake basins, merge. One series continues westward and stretches to Crow Hill in the southwest. The other heads southeast toward Horseshoe Lake, passing Devil's Heart Butte. These two have patches of intermittent ice-shoved hills. To the south of the ice-shoved hills is a melt-water drainage system. Melt water from the glacier at the ice-pushed hills was transported by all channels into the

Sheyenne valley, which is located to the south. It isA subglacial melt-water may head toward the Devils Lake basin to the north. These bays and valleys are assumed to be tunnel valleys created beneath the snow by melt-water erosion. Sullys Hill itself is traversed by narrow tunnel valleys that run straight to open spillway channels to the south. The fact that this group of glacial landforms has been well preserved indicates that either glaciers did not pass through this area or they did not significantly alter the hill-hole pairs and related melt-water formations. Large ice-shoved hills, melt-water spillways, hydrodynamic deposits, and tunnel valleys are all associated with the idea that a significant amount of water was discharged during the formation of these landforms. This water came from the Spiritwood aquifer as well as the glacier.

3.7 GLACIATION

In areas where sedimentary layers are either weakly consolidated or unconsolidated, hill-hole pairs are commonly found. In depths of 200–300 meters, some researchers observed hill-hole pairs close to the Norwegian continental shelf (Bellwald et al., 2023). Glaciation in this context was undoubtedly marine-based. Bathymetric maps, seismic profiles, and sediment samples are used. Horseshoe hill is more than 100 meters above the surrounding sea floor, which is roughly 300 meters below sea level. The hill's southern edge bends to the southeast into a lower ridge, while the hill's northern edge is a sharp escarpment that stretches eastward into a low ridge. The hill has an area of about 20 km_2, and it is positioned on a substratum made of Quaternary material. The hill crest sediment core has a shear strength of 10-90 kN/m_2. The Horseshoe basin is estimated to be 0.7 km_3 in volume and to cover an area of around 35 km_2. The Horseshoe and the nearby sea-floor depression are a pair of practically perfect hill-holes that have been maintained. The hill border with the undisturbed sea bottom and its steep distal (NW) margin suggest that the hill was not covered by the glacier margin after deformation. Thus, the majority of the hill is thought to have resulted from pushing glaciers. This progress manifested as an ice tongue. The formation of the hill's lower limbs could have resulted from either lateral squeezing along the material of an ice tongue being scraped away during the ice tongue movement. Fast ice movement may be accompanied by a surge and rapid sediment deformation.

Steinbitryggen is an uneven ridge on Fuglcybanken that rises to a height of roughly 90 meters above sea level. The ridge has a volume of about 1.6 km^3, extends about 30 km east-west, and covers an area of around 58 km^2. A depression known as Sopphola sits at the easternmost point of Steinbitryggen. It is 56 km^2 in size and has a volume of roughly 1.7 km^3. Steinbitryggen is situated next to Sopphola and has a triangle shape. It also has multiple ridges.

Sopphola is made up of multiple distinct basins. Glacier is likely advancing from the southeast to the northwest based on the Sopphola terrain and Steinbitryggen. It forms a low ridge that stretches more than 20 kilometers west. The thickness of the quaternary layers in the area may be more than 250–300 meters (Derbyshire et al., 1991)including data on glacial and periglacial geomorphology, sedimentology, tectonics, stratigraphy and lake history, have been used to reconstruct ice margins and palaeosnowlines. Three, and in places, four distinct glaciations have been recognised, the most extensive occurring in the later Middle Pleistocene. Valley and piedmont glacial systems, with some mountain ice caps and a small ice sheet on the upper reaches of the Huang He, have been recognised but there is no evidence of a single Tibetan ice sheet. As the Plateau and the Himalaya-Karakoram underwent accelerating uplift through the Quaternary, there occurred progressive desiccation in the interior as the influx of Indian Ocean moisture was constrained. Equilibrium line elevations in the last glacial maximum were 4000 m in the south, east and northeast Plateau margins, rising to 5500 m in the northwest interior. The heat island provided by the Plateau, the progressive reduction in precipitation most marked during the glacials, and the strong southeast to northwest precipitation gradients, produced a glacier distribution pattern dominated by the trans-Tibetan mountain ranges. Tectonics and periglacial (including aeolian. The topographic ridges in the adjoining part of Steinbitryggen and the many basins within Sopphola imply that whole blocks were taken out of the source basin and piled into the ice-shoved hill. This further implies a distinct decollement along which displacement occurred and a comparatively high shear strength of the displaced material. The structure was pushed by glacial ice some tens of kilometers after Sopphola and the hill's initial thrusting. Nevertheless, whether the original pushing occurred subglacially or at the ice margin is unknown. There has been minimal post-glacial erosion or sediment cover to alter the original topographic character. These formations show no signs of postglacial changes. Based on this, researchers deduced that glaciotectonism was young, most likely dating from the Late Weichselian era.

3.8 COMPOSITE RIDGES

Composite ridges can form in a variety of sizes. Long (50 km wide/200 m high) are possible dimensions for large composite ridges. Composite ridges are frequently concave, arcuate upglaciers with a 2 to 10 km radius. Individual ridges are usually a few hundred meters wide and show several meters of topographic relief. The curve of the ice shoved hill is followed by a subparallel pattern formed by ridges and intervening valleys, which frequently contain small lakes. The edge of the tongue or lobe of ice that protruded up the ridges is indicated by this arcuate pattern. Topography and structural characteristics are closely related. Large composite ridges, which are a substantial portion of the ridges, typically require significant disruption of pre-Quaternary bedrock. But typically, structural disturbances only reach a depth of 200 meters. Perhaps the most prevalent glaciotectonic landforms are small composite ridges (less than 50 meters high). They share the same structural and morphologic characteristics as massive composite ridges. Similar geographic features like escarpments, islands, or valleys also contain small composite ridges. These kinds of ridges are unable to sustain a high topographic relief because they are more vulnerable to erosion from glaciers and other sources. Ice margins are linked to both large and small composite ridges. Typically, the thrust belts and folds that make up composite ridges are separated and stacked to create an imbricated structure. Because of this, composite ridges are allochthonous in the glaciotectonic terminology, and it might be able to identify the upglacier depression that provided the material for the ridges. But often, as with the hill-hole pair, it is impossible to pinpoint a precise discrete source basin.

Figure 3.7 Prophets Mountains

Source: https://commons.wikimedia.org/wiki/File:Prophet_Elias_Mountain_%284427796428%29.jpg, Creative Commons License.

Located on the Missouri Coteau in central North Dakota, Prophets Mountains is a suitable example to explain massive composite ridges (Aber et al., 1989). The Prophets Mountains span about 20 km^2 and have a maximum elevation of 690 m. The Prophets Mountains are made up of several parallel ridges that primarily run north-south. The hill is asymmetrical generally, with steeper slopes and higher altitudes on the western side. Over 100 meters of relief is present overall, with individual ridges rising 10 to 25 meters above neighboring valleys. Most of the area around the Prophets Mountains is covered in till, which is mostly left behind by glaciers. Along the northwest of the Prophets Mountains, a sequence of esker ridges has formed.

At the southeast border of the Prophets Mountains, (Figure 3.7) bedrock that has been folded and is at least 100 meters above its typical stratigraphic location. The Prophets Mountains' ridge shape reflects ice pushing by an ice tongue moving eastward. A bedrock depression measuring approximately 75 km^2 and up to 50 m below the preglacial valley bed has been found during drilling (Pomper et al., 2017). The Prophets Mountains were formed on a large, intricate aquifer that was formed in two places: partially in a flat valley that is a tributary of the preglacial Knife River, and partly in Cretaceous sandstone. Currently, this aquifer allows groundwater to drain eastward, but during the pushing of the Prophets Mountains, glacier ice obstructed this flow. According to Bluemle and Clayton, ground-water flow was reversed and water pressure increased to the point where glacier thrusting was made possible. When referring to ridges formed by shoving ice, the term 'push moraine' is often used. Push moraines refer to a group of composite ridges that are partially or entirely composed of glaciogenic strata. It is inappropriate to refer to composite ridges that have a significant amount of non-glacial material as push moraines. Pushing of concurrent proglacial or ice-contact drift during continuous glacier progress is one typical scenario for push moraines. During the Little Ice Age, the Arctic islands in northeastern Canada saw the creation of several push moraines of this kind in outwash deposits due to glacier advances.

3.9 BEDROCK

The Dirt Hills and Cactus Hills have distorted features of upper Cretaceous bedrock groups. With a total thickness of up to 90 meters, the deformed layers are primarily terrestrial sandstone that is either bentonitic or kaolinitic, together with mudstone and lignite. The Bearpaw Formation's unbroken marine shale lies beneath them. The general structure of the Cactus Hills and Dirt Hills is made up of thrust and folded faults arranged in an imbricated pattern. Drift covers the bedrock at most places, although it is not a factor in the deformations. In the pebble fraction, dolostone and crystalline erratics dominate the drift. Nevertheless, due to glaciotectonic uplift, older drift blocks rich in chert and quartzite are found at some sites. Displacement has not only happened horizontally but also vertically. The finding that ice-shoved ridges on the Missouri Coteau are direct or first-order morphologic representations of bedrock structures created by ice pushing. The majority of bedrock formations are associated with a single ice pushing episode and direction, although others exhibit evidence of many episodes of deformation. Larger folds and fault blocks are made up of thicker sandstone layers; thrust faults are typically found inside lignite or claystone beds. Since ridges are first-order morphologic representations of bedrock features, topography can be used to determine the main regional structure of the Cactus Hills and Dirt Hills. Aerial photos show the ridge form clearly because of the relatively dry climate and limited vegetation. Class I ridges were most likely formed by a previous, unidentified glacier push. The final ice advance pushed up Class II and III ridges to push onto the Missouri Coteau(Bennett, 2001).

CHAPTER 4

ENVIRONMENTAL CHANGES AND GLACIERS

Contents

4.1 INTRODUCTION

Glaciologists have long believed that glaciers flowed over hard, lithified, impermeable beds until the last 25 years (Ravier & Buoncristiani, 2018). Hence, any décollement at the point between the snow and the bed must occur from ice sliding over a hard rock surface. Weertman proposed that the basal drag must be produced by movement due to plastic flow and pressure melting/regelation around obstacles on the bed and that sliding at the ice-bed interface must be accompanied by an intervening thin water layer, which would prevent tangential friction. Direct field observation has validated a large portion of this conceptual framework. Nonetheless, experimental stress measurements carried out beneath a glacier demonstrated that debris trapped in basal ice supplied a significant component of the drag at the glacier base. Water from basal melting is thought to move in a thin layer between the ice and the bed until it can enter subglacial drainage channels. It was demonstrated that a significant amount of the forward movement of a glacier might be attributed to the shear deformation of the soft, unlithified sediments beneath a glacier. Fischer and Clarke noted that at times of high-water pressure, slide/drag occurred at the ice-sediment surface, but during times of low water pressure, a shear movement was localized in the sediments. The effective pressure at the ice-bed interface may be affected by groundwater flow. Additionally, some researchers demonstrated that a substantial amount of basal meltwater may be released in permeable subglacial beds as opposed to being limited to flowing through channels at the ice-bed interface. Effective pressures, the threshold for deformation in unlithified subglacial sediments, and changes in the position of décollement between the ice-bed interface and underlying sediments are all influenced by the intergranular drainage process.

4.2 SEISMIC SURVEY STUDIED

Our knowledge of the nature of the basal boundary condition and the sedimentary processes at the ice-bed interface is based on some discoveries. Seismic surveys and drilling in the vicinity of several Antarctic ice streams have revealed an underlying soft bed, although we are currently only able to evaluate the nature of beds beneath the ice sheets of Greenland and Antarctica over a very small area. These ice streams are thought to be involved in the fast flow process because they flow at high speeds under driving loads as little as 10 kPa (Payne & Dongelmans, 1997). The characteristics of these beds have a major influence on a wide range of large-scale dynamics of ice sheet behavior. The function of the bed in connection to the essential hydraulic processes that establish the effective pressure regime and the type of frictional interaction between the glacier and bed were examined. The rate at which water recharges into the subglacial system and the efficiency with which it drains from it regulates the water pressure at the base of a glacier. Minimal water pressure and relatively low water pressure gradients are linked to efficient drainage. High water pressures and relatively high-water pressure gradients are linked to ineffective drainage. The amount of channel spacing that would be required would decrease to the order of tens of meters if large amounts of summer meltwater were to reach the bed from the glacier surface. An annual cycle of channel development is therefore very likely to result from an excessive oscillation of the system between winter and summer. According to Boulton, the winter channels beneath an ice sheet are long-lasting. The closely spaced network of summer channels may represent the slowly moving, high-water-pressure canals. Because these channels would not last long, they would not be able to unite and form an effective system of low-pressure channels that could draw down water pressures over large areas as well-developed R-channels.

4.3 BEDROCK

According to Boulton and Dobbie, the water pressure gradient in a till covering low permeability bedrock tends to be horizontal and is directed toward significant low-pressure R-channels that serve as groundwater flow sinks/reservoirs. However, there may be a high, downward water pressure gradient, powerful enough to liquefy the till and force it into bedrock crevices, where there is a well-connected fault system in bedrock, such that water flow along interconnected faults is able to draw down water pressures. The outcomes of this process are frequently observed in bedrock regions, where till dykes can slice through bedrock for many meters. Ice–bed interface can freeze when ice streams stagnate, potentially leading to processes of sediment inclusion via freezing. Evidence from northwest Greenland indicates that a frozen sediment deposit can remain mostly undisturbed by the glacier sole, and similar findings were found on a subglacial rock bed in Antarctica. The adhesive strength of an ice–rock interface at temperatures well below the melting point was significantly greater than the maximum average shear stress that is typically generated at the base of an ice sheet. With basal shear forces of the order of 100 kPa, we can anticipate that internal flow will predominate in the movement of ice sheets and that sliding will be restricted at the ice–bed interface (Ottesen et al., 2008). However, sliding can also occur at temperatures below the melting point. It can also abrade bedrock, which can result in debris assimilation in basal ice and deform any frozen subglacial sediments through deformation. While substantial blocks of debris or sediment may be torn away from the substratum by local stress, cold-bed erosion rates are generally expected to be quite minimal.

4.4 PRESSURE EFFECTS

If obstructed, huge frozen silt and rock masses can be transported by ice. High water pressure is caused by inadequate drainage under a subglacial frozen basin. High water pressures at the base can lower the effective pressure and friction to a point where sliding can happen along the frozen–unfrozen interface, assuming good adhesion between the glacier sole and underlying frozen materials. There have been recommendations that this process could allow very large rocks or sediments to move with the glacier. It could also explain many of the large glacier tectonic structures as well as large, relatively undisturbed pre-existing rock or sediments found in glacial tills. With the basal décollement located below rather than above the glacier, permafrost essentially becomes a component of the glacier. Surface runoff enters a wide subglacial drainage system early in the melt season. A hydraulically effective network of subglacial channels is extended later in the season when the surface snow in the ablation area is removed, leading to higher diurnal runoff cycles. Drumlins and other subglacial formations are frequently regarded as concealment due to the lack of observation regarding their development.

Figure 4.1 Drumlins

Source: https://commons.wikimedia.org/wiki/File:Drumlin_landscape_-_geograph.org.uk_-_1327989.jpg,Creative Commons License.

Drumlins (Figure 4.1) are formed by direct ice action on a deforming bed. Turbulent, low-viscosity flows that behave differently from ice-induced high-viscosity flows and deforming sediment straining under ice-induced stress result in eroded structures. It is highly improbable that wind and fluvial processes may be employed as equivalents for ice action and vice versa. It is critical to show that meltwater was present at the initial formation site of drumlins and other bedforms if the meltwater explanation for these features is to be accepted. According to some models of the formation of ice sheets, there are cold beds that rule out the meltwater concept. The sediments and geomorphology of the inner zones of the continental ice sheets provide evidence for meltwater, and the lack of melting would limit the potential for hydraulic interaction between these inner and outer zones. The reconstruction of the processes that produced the sediment and landforms yields the thermal and liquid water conditions. The water required for their creation is triggered by viewing them as meltwater formations. Whether or not there was enough meltwater to create the subglacial landscape is the most important question. Reservoir storage is needed to accommodate the necessary amount of meltwater. Both the water source and the reservoirs could have been supraglacial or subglacial. A surplus of subglacial water may favor supraglacial reservoirs. The presence of reservoirs at the glacier bed could result in surface deformation and the possibility of supraglacial reservoirs in a significantly larger ablation zone as a result of climate change. Megafloods may have been caused due to a supraglacial lake with the bed (Carling, 2013).

The stratigraphy in the inner zone directly indicates meltwater at the ice-sheet bed. The striated bedrock of Lake Mistassini and Lake Albanel is the underlying layer of lodgment or maybe deformation till. Melting was necessary for the tills to be deposited. Abundant meltwater is necessary for entrainment and deposition for basal debris to melt out after first freezing. Similar in texture to tills, Rogen moraine deposits have larger strata of sediment that contain even more water-sorted deposits. There is no question that a large portion of the interior zone of the Laurentide Ice Sheets was warm based on sedimentological and geomorphological data. Similar striations, lodgment tills, and Rogen ridges can be found inside the Scandinavian Ice Sheet. These features are made of extremely localized lithology with widely distributed sediment, which suggests that the sorted sediment was either entrained before melt-out or formed by erosion and deposition of meltwater. Meltwater was necessary at the bed for the so-called Sveg tills to facilitate the deposition of debris. It is not essential to conduct a detailed investigation into the genesis of Rogen moraine or Sveg tills in order to bolster the theory that the interior of the Scandinavian and Laurentide ice sheets had an abundance of meltwater. The subglacial silt contains significant amounts of meltwater in the inner zone. The landscape that the Pleistocene ice sheets left is one of their most attractive features. End moraines are remarkably rare along the flow channels; these subglacial landscapes were developed around the period of the Last Glacial Maximum (LGM) of the ice sheets (Hodgson et al., 2014). This suggests that a substantial portion of the deglaciated landscape did not experience the formation of recessional landforms. There was not enough slope to maintain glacier movement when the ice reverted to its bed and sloped back up-glacier. The Laurentide ice sheet bed had a variety of bedforms. The subglacial processes appear to have stopped abruptly over a large portion of the area. Meltwater is primarily responsible for creating the landforms. Not all subglacial landforms can be classified as fluvioglacial.

4.5 VARIATIONS AND FACTORS

Changes in glacial mass balance, which are connected to variations in ocean and atmosphere temperatures, are primarily responsible for variations in glacier extent on decadal to century timescales, such as the Little Ice Age (LIA). Non-climatic factors, like glacial isostatic depression of the crust, have an impact on changes in glacier or ice-sheet extent on the Milankovitch or orbital cycles. These factors can impact ice-sheet mass balance, particularly if they end in the oceans. Moreover, the biosphere's involvement must be an essential component of the entire system in light of recent evidence on variations in the greenhouse gases CO_2 and CH_4 during multiple glacial cycles. In fact, some contend that changes in CO_2 are either a key amplifier or the main cause of climate change. Our understanding of the relative relevance of the various forcing functions as they fluctuate temporally and spatially might be greatly expanded by modeling the reaction of ice sheets to different climatic scenarios. Since the cryosphere is a crucial component of the climate system, it is appropriate to emphasize how glaciers react to variations in the biosphere, atmosphere, and ocean climate. The extent, thickness, and shape of the massive ice caps and ice sheets also have a positive feed-back relationship, particularly with respect to the atmospheric circulation. At the most basic level, variations in the amount of snow and ice covering a region of the LIS would influence the net radiation budget and the surface reflectivity, while variations in the height of ice sheets could potentially impact the upper atmosphere. The significance of ice sheets on the climate during glacial intervals is assessed using global circulation models (GCMs). The effect of glacier meltwater on ocean circulation is another important feedback. It is hypothesized that the massive outburst of glacial meltwater from the retreating Laurentide Ice Sheet (LIS) into the North Atlantic will reduce thermohaline circulation (THC) and cause sudden cold spells (Wiersma & Renssen, 2006). However, the effect of freshwater during periods of glacial build-up has received less attention. One would be obliged to believe that freshwater entering Hudson Bay and the Labrador Sea as well as the quantity reaching the Arctic Ocean would decrease as the ice sheets grew throughout the

Northern Hemisphere. Furthermore, when the amount of ice on Earth grew, the amount of relatively fresh Pacific water entering the Arctic Ocean decreased and eventually stopped when the sea level dropped to roughly -50 meters. The retreat or advance of glaciers is commonly associated by glacial geologists and geomorphologists with particular climatic parameters, particularly summer temperature. For instance, the location of the equilibrium line altitude (ELA), which depends on the net mass balance, is connected to the overall energy balance. It is evident that an ice mass archives the regional climate through a variety of factors. The energy balance on the glacier surface determines the critical mass balances that affect a glacier on a yearly or longer cycle. For the majority of the glaciers, or those with surface characteristics classified as either subpolar or polar, the balance of net radiation, rather than temperature, is the critical issue in summer losses. The mass loss that results from specific summer weather and the glacier forward motion, which is the integrated flow response to events in the accumulation and ablation zones over a certain amount of time, determine the position of the glacier frontal position year-to-year.

When glaciers end in marine or lacustrine basins, the scenario is very different. The mass balance and resultant glacier reaction are influenced by a non-climatic factor. It must be acknowledged that the factors affecting calving are poorly understood. The common effect on glaciers or ice streams that end in freshwater or saltwater is an additional mass loss to the glacier system by the mechanical removal of icebergs by calving. Nonetheless, research conducted in Antarctica has demonstrated that mass can be added through the freezing of seawater at the ice shelf base. To balance the surface freshwater outflow from the summer ablation, it was observed that the ocean dynamics in Kangerlussuaq Fjord caused the importation of relatively warm, salty water in East Greenland. Thus, the oceanic estuarine circulation and the atmospheric climate are positively correlated. The importation of warmer waters from the deep shelf, which is eventually fed by Atlantic Water in the Irminger Current, would be diminished if, for instance, cold summers produced less meltwater. This would then reduce basal melting at the ice front. Tidewater glaciers may rarely move through fjords with ocean depths ranging from 100 to 1000 meters during periods of positive mass balance (Syvitski, 1989). There is a debate about how outlet glaciers develop along deep-water fjords because there is a propensity for calving to increase as the ice advances into deeper water. A shoal (moraine) is created by the advancing ice moving over fjord sediments to the ice front (temperate Alaskan fjords).

4.6 EFFECT OF CLIMATE

Earth scientists study sedimentary records for glacier response to climate both on land and in the sea. Records must be interpreted in a way that makes it apparent what is known or presumed about depositional mechanisms. Consider the question of dating moraines from the Canadian Arctic subpolar regions vs. temperate glacier forelands of Norway. In the first scenario, more materials from rockfall are added to the sediments entrained in the lower traction zone, resulting in the formation of moraines. Every year, during the summer, when ablation is at its highest, the margins slightly recede, and during the winter, when surface melting does not counteract the forward motion, the margins readvance.

Figure 4.2 Lichens

Source: https://commons.wikimedia.org/wiki/File:A_glacial_erratic_%28boulder%29,_covered_in_lichen_and_moss,_on_the_slopes_of_Moel-y-Ci.jpg, Creative Commons License.

Lichens (Figure 4.2) may grow on surface stones in moraines that are isolated due to summer ablation. In contrast, sediments are carried to the surface of the ice sheet in the subpolar landscape, where they undergo a process of sediment redistribution that results in ice-cored moraines. The relationships between the formation of moraines and climate in such an environment are even weaker than in the case of temperate glaciers. The established ages of the moraines are periods of ice-core stabilization, which delayed the timing of climate change by several decades or even centuries. Without discussing the role of swelling glaciers and the driving of glacier reaction by non-climatic influences, the discussion of glaciers and climate would not be comprehensive. Responses from surging glaciers and glaciers ending in a lake or marine waters may deviate from trends in climate. The margin shifted from a calving-dominated area to one where summer conditions forced mass loss, and large moraines associated with the LIS could form due to the reduction in the ice sheet mass balance. Most glacierized regions of the world experience surges in their glaciers, which are characterized by long periods of slow fall. They can be distinguished by their folded medial moraines, although It isuncertain if previous surges left any recognizable deposits or depositional architecture(Benediktsson et al., 2008).

The upper limit of lateral moraines of a specific age can be used to define the ELA on alpine glaciers, both past and current. Many methods have been used to measure changes in the elevation of the ELA throughout time. The delineation of the glacial threshold (GT) or limit is a technique for identifying the regional climatic restrictions on the distribution of ice bodies, while the ELA is glacier-specific. The impact of winter accumulation on a glacier mass balance is not taken into consideration by this method, which often expresses changes in an ELA or GT elevation as variations in temperature. Relatively few attempts have been made to connect glaciological climatic indices. Reiterating that the mass balance is the result of a winter accumulation less a summer mass loss is crucial. Therefore, although assumed temperature changes are frequently the focus of work on past glacial changes, variations in accumulation must also be considered. Similarly, ice-rafted debris (IRD) is being used more and more as a substitution for Holocene climate change in marine sediments. The movement and deposition of sand-size particles from icebergs or from material entrained in sea ice is the first question to be answered by the IRD event. The increased IRD at a location could indicate that icebergs have redirected or that sediment entrainment in outlet glaciers has altered. Most people assume that changes in the iceberg flux, which are linked to climate change, are the only causes of variations in IRD.

4.7 SALINITY

The Great Salinity Anomaly (GSA) and other changes in marine conditions, such as the amount of sea ice and variations in the thermohaline circulation, have a significant impact on Iceland, making it an important region in the North Atlantic (Belkin et al., 1998). According to many studies, midlatitude glaciers have rapidly receded over the last two to three decades, when scientists have begun to focus on the impact of human activity on climate. There is a clear correlation between climate, sea level fluctuations, and glacier recession; nevertheless, due to the massive scale of Greenland and Antarctic ice sheets, it is challenging to assess the impact of climate change on their mass balance. If an increase in winter temperatures is the apparent result of global warming, then a logical consequence would be an increase in snow accumulation at high latitudes, which could lead to a positive mass balance in the Greenland or Antarctic ice sheets. Furthermore, over the last two to three decades, there has been an increase in the discharge of the major Russian rivers that drain into the Arctic Ocean, contributing around 1400 $km^3\ yr^{-1}$ of freshwater to this ocean (Gordeev, 2006). This freshwater, which is exported from the Arctic Ocean as sea ice, has affected the general climate as well as the rate of convection, or the thermohaline circulation, in the Arctic Ocean and the Nordic Seas.

With the increased export of freshwater into the Nordic Seas, mostly in the form of sea ice, and its effect on the thermohaline circulation, the idea of a warming, especially at high northern latitudes, carries with it the seeds of a significant negative feed-back loop. The export of this freshwater could have a disastrous effect on the production of North Atlantic Deep Water (NADW). A significant drop in temperature would occur over northwest Europe if the ocean conveyor system were to shut down due to a decrease in the North Atlantic current. An early example of this kind of event was the GSA in the late 1960s. The concerns about the collapse of the West Antarctic Ice Sheets, which were a subject of ongoing debate beginning in the 1960s and 1970s, are more than critical scenarios if the climate is warming and sea levels are rising. Studies of variations in the length of 27 non-surging glaciers in Iceland provide a regional example, demonstrating that the advance/retreat records of non-surging glaciers show a clear relationship to climate. Following a period of cooling that began in the 1940s and concluded in a significant shift in the mid-1960s, around half of the glaciers have begun to advance again.

4.8 TIMESCALES

End moraines, which are found downstream from modern glaciers, serve as the main source of evidence for the glacial reaction to climatic perturbations throughout the LIA period, or around the last 600 years. The literature suggests that the LIA was well begun by the 13th and 14th centuries. Compared to the response and interactions on longer time-scales, the boundary conditions for LIA glacier-climate systems are less complex. The quantity of tidewater glaciers with their unique dynamics are more noticeable on longer timescales, especially throughout the Holocene. The dynamics of change on LIA timescales are mostly confined to the atmospheric climate system, which includes the oceans. Small-scale tectonic motions, such as glacial isostatic depression, are likely present on these time intervals but do not induce a glacier reaction. Relative sea level (RSL) variations are likewise negligible (Rovere et al., 2016). The Greenland isotopic records spanning from 1400 to 1980 may exhibit a negative correlation with temperature anomalies in Europe due to the changing nature of the climate. Some researchers assessed the interactions between glacial history and the surrounding land/oceans in terms of the feedback related to the GSA of the late 1960s. While there is disagreement regarding the exact cause of the salinity anomaly in the Nordic Seas, it is widely accepted that the anomaly most likely began thousands of kilometers away inside the Arctic Ocean borders, either as a result of an anomaly in the McKenzie River flow or from an increase in sea ice transport through Fram Strait brought on by atmospheric processes. At least a portion of the climate system exhibits some degree of self-regulation. It is believed that hydrological changes on the Canadian Arctic border spread to the Greenland and Iceland areas, where they caused the surface waters to become fresher. This lowers cyclogenesis, which in turn lowers precipitation and lessens the freshwater anomaly. Changes in the global ice volume and adjustment are linked, over longer timescales, to variations in the bathymetry of the major Canadian Arctic Channel sills. As a result, an increase in sea ice is linked to a drop in temperature, a drop in precipitation, a fall in the ELA, and an increase in glacier extent. Variations in sea ice also affect temperature and precipitation over northern Iceland. The decrease in temperature is related to the increase in sea ice off the coast, and is more significant than the fall in precipitation, notwithstanding the lack of long-term mass balance measurements in northern Iceland. Around 570 meters above sea level, the basalt surface is home to the Drangajokull glacier. The ages of LIA moraines in Iceland show significant stretches of moraine formation over the previous three centuries, however, there is also evidence of glacial moraine deposition approximately 700 years ago.

Around 1450 AD, there was a cold period that lasted for a while. Other extended periods of cooler weather occur around 1600 AD, 1700 AD, and roughly between 1800 and 1900 AD. The information from Stykkisholmur, in the northwest of Iceland, dates back to 1820 AD(Axford et al., 2009). The summer and winter trends showed an increase in temperatures, led by winter values, beginning around 1900 AD, after a period of cold winter and summer temperatures placed around 1860 AD. The association between winter and summer temperatures is clearly visible. Based on the Greenland ice-core data, a reconstruction of the Iceland Low-pressure system evolution has been studied. This reconstruction indicates that the Iceland Low intensified and that a major change had taken place at approximately 1400 AD, when general high-pressure conditions abruptly changed to lower pressures. This would suggest that winter accumulation over Iceland would have grown between 1400 and 1450 AD, based on the most recent data from Stykkisholmur. Iceland is a small island of about 100,000 km^2, but research on its climate indicates that weather stations there strongly correlate with one another. Given the strong correlation between the 50-year time series from the Siglunes hydrographic CDT data, off the coast of north Iceland, it is logical to assume that changes in the marine area can be used to predict land temperature conditions. The stable $\delta 18O$ of foraminifera serves as a substitution for marine temperatures during the late Holocene since research has demonstrated that these predominantly reflect temperature variations.

4.9 GLACIAL RECORDS

The early and mid-Holocene climate is currently under scrutiny; comparisons between the Holocene and the atmospheric greenhouse gas interglacial records found in ice cores imply that both the CO2 and methane records reflect human habitation. For the last 12,000 years, the climate component of the glacier relationship is based on a range of data from ice, lake, and marine cores (Larsen et al., 2012). The glacier response is frequently recorded in a sequence of terminal moraines at different distances from the current glacier margins, alterations in the historical ELAs of glaciers, or fluctuations in ice-rafted debris (IRD). Therefore, glacial activity is also inferred from changes in sediment type, such as grain-size, in both lake and marine environments. Climate data has been established that pertains to changes in glacier mass balances during the previous 12,000 years. The most accurate estimates come from ice cores, where variations in accumulation have been determined; stable isotopes of precipitation, which are thought to primarily reflect air temperature; and well temperatures, which represent a long-term integrated measure of air temperatures at roughly 3 km over central Greenland. From periods of moraine formation, inferences about the climate of the neoglacial period over Iceland have been made. These may underestimate the actual number of glacier advances due to older moraine systems. The seafloor temperature data indicate significant temperature fluctuations, particularly within the last 6 cal. Kyr BP, and a peak in heat about 7 cal. Kyr BP (Hald et al., 2007).

It is challenging to demonstrate that the Holocene IRD sediments in the North Atlantic have a glacial origin as opposed to the competing theory that they are made of sand-sized particles carried on or inside sea ice. It is unclear why the delivery and transport of Icelandic volcanic debris to sites to the south and east of Iceland is not primarily a wind-driven process that is independent of sea ice as a transport agent. There is growing evidence that Icelandic tephras can be found in the Greenland ice core, sediments on the east Greenland shelf, lakes in Ireland, Britain, Norway, and Sweden. It is necessary to conduct studies on the constant flow of Icelandic volcanic debris at various locations throughout Iceland to account for the variability in tephra generation while estimating the potential contribution of IRD. The evidence for global meltwater surges associated with late marine isotope stages (MIS) 2 and 1 is derived from relative sea-level records from sites that are far from ice sheets. This evidence differs from the evidence for meltwater events derived from the planktonic isotopic evidence of foraminifera ocean sediment core records for the northern North Atlantic. The former displays massive meltwater bursts linked to sharp increases in sea levels worldwide. Given the abundance of information available for the deglaciation period (19–10 cal. kyr), it is interesting that no research has been able to identify the ice sheet(s) associated with MWP1a. Although the pulse-like nature of the process suggests other processes, such as calving into pro-glacial lakes, must have occasionally predominated, the global meltwater pulses are most likely caused by increased melting on the ice sheets linked to the rise in summer insolation. Planktonic isotope data from the northern North Atlantic suggests that surface melting on the Greenland, and LIS ice sheets occurred before MWP1a. This could be due to the rapid retreat and collapse of ice across glacierized continental margins.

4.10 MARINE SEDIMENTS

From the 1970s onward, studies on marine sediment cores have demonstrated that variations in the isotopic alterations of foraminiferal δ18O correspond with those expected from calculations of variations in Earth's orbit (Bol'shakov, 2014). It was generally accepted that variations in solar insolation at high northern latitudes forced changes in the ice sheets. In particular, peaks in spectral analysis with periodicities of approximately 22, 41, and 100 kyr match variations in obliquity, precession, and eccentricity. The sudden halts to the glacial cycles imply that positive feedback is provided during the deglaciation processes by sources other than the steady variations in solar insolation determined by orbit. Many continental ice-sheets encountered tidewater during glacial phases, and the collapse of the marine-based ice sheets in the Northern Hemisphere provides one explanation for the abrupt halts. Upon examining glacier-climate interactions across 104-year timescales, researchers addressed the expansion and contraction of massive ice sheets. On these time-scales, changes in the elevation of ice sheets and relative sea level will occur at the margins of ice sheets as the time for the process of glacial isostatic adjustment is of the order of 2-3 kyr for the LIS. The interaction between the height and albedo of large ice sheets and the climate system is also significant, as is the impact of freshwater inputs to the ocean and the effect on the THC. The vast Northern Hemisphere ice sheets that cover Fennoscandia and northern North America are marine-based because they are centered spatially over the vast inland seas of the Baltic/ Gulf of Bothnia and the Foxe Basin/Hudson Bay. A significant portion of these ice sheet beds were deep below sea level at glacial maximums due to the weight of the sheets and the resultant glacial isostatic depression of the crust.

The West Antarctic Ice Sheetis sensitive to both historical and present climate change theoretically. It was unexpected, though, that the evidence found through studies of marine cores in the North Atlantic for sudden and massive changes in ice-sheet discharge was linked to periodic collapses of the LIS, known as Heinrich or H-events. The nature of the glaciological response did not come into focus until 1992. Heinrich proposed a link with radiation forcing, which was the prevailing paradigm at the time. As a result, the focus of research on the relationship between the ice sheet and temperature changed in the 1990s to consider evidence of abrupt changes in the cryosphere and climate system at far greater periodicities than changes in solar insolation caused by variations in Earth's orbit around the sun. The results of the study of marine sediments have dominated the research works on glacial occurrences dated between approximately 10 and 50 cal. kyr for the past ten or more years. There isn't much proof that glacial H- or D-O (Dansgaard–Oeschger) events occurred. The marine studies revealed evidence of major meltwater inputs to the ocean as well as other glacial reactions, most of which were understood in terms of changes in the ice rafted component (IRD) of deep-sea sediments. Major glacial cycles caused the ice sheets to thicken and spread. This has caused ice sheets to extend onto the continental shelf in many places. The relative sea level at the offshore margin of the ice sheet will likely be rising during the advance phase of the glacial cycle due to the crust experiencing glacial isostatic depression. This is dependent on the balance between regional ice-sheet growth, which causes isostatic depression, and global ice-sheet growth, which extracts water from the ocean.

4.11 DATING

A buoyant force is pushing the ice sheet away from its bed because of the variations in density between ice and marine water(Hughes, 2003). This reduces friction at the bed and speeds up the flow of glaciers, which cannot be counterbalanced by an increase in accumulation. As a result, ice margins on glaciated continental margins collapse and recede quickly. Marine upheaval at the sea floor would result from glacial isostatic adjustment as the glacier recedes. There are thus indications of a self-regulating cycle of glacial response, whereby an increase in the regional mass balance leads to an advance of ice onto the continental margin; however, rapid calving triggered by variations in water depth may subsequently cause the termination/end of the advance and the retreat. These responses are essentially distinct from the dynamics between climate, mass and balance that govern activity on land. The relationship between a particular glacier or ice sheet reaction to climatic forcings becomes increasingly obscure as we go further back in time. In fact, the history of Quaternary glacial cycles is based on shifts in the isotopic composition of foraminifera, which reflect the global hydrological balance, rather than on glacial geological evidence per se. Following World War II, large-scale marine expeditions became common, and with the advancement of 14C dating, δ18O measurements of foraminifera, and paleomagnetic chronology, it was eventually realized that the global hydrological balance, as it related to the growth and retreat of ice sheets, shared similarities with the estimated variations in solar insolation linked to orbital or Milankovitch variations. Even though the correlation is generally acknowledged, it is still unknown how such significant variations in global ice volume are caused by the very minute variations in insolation/radiation. The glacier–climate system needs a few non-linear feed-backs, one of which is the feed-back between variations in the amount of snow cover due to its high albedo. The importance of the biosphere is growing in light of recent evidence on the phasing/dating of ice volume and greenhouse gas content over the previous 0.5 million years. The swift mass loss of those ice-sheet borders, which extended onto the continental shelves, may provide the best explanation for the abrupt terminations/ends. Rapid deglaciation and a positive feedback loop from the melting of ice sheets and subsequent rise in sea level could be the outcome of this process.

The primary transition that took place on Quaternary time-scales was the change, which happened approximately 0.7–0.8 Ma, from the 42 kyr periodicity that was previously dominant to the 100 kyr periodicity that followed. The most recent evaluation of the contribution of solar insolation to orbital-scale climatic change was given by Ruddiman. Though it does not directly address the question of the histories of individual ice sheets, the shift in stable oxygen isotopes reflects an integrated global assessment of the addition and removal of ice on land, and it is certainly unable to resolve the record of mountain glaciations, other than by inference. Although it is unclear how precisely the history of this ice sheet links to the significant variations in δ18O, one would think that on Quaternary time-scales the LIS is still the dominant component of the δ18O signal in foraminifera. Ruddiman developed a model to explain how changes in CO_2 and CH_4 levels and variations in insolation would affect the glaciers using the greenhouse gas data from the Greenland and Antarctic sediment cores (Ruddiman, 2003). The understanding that ice-sheet growth and decay delay the forcing functions is acknowledged. Moreover, the focus is mostly on the significance of summer insolation, particularly in the middle of July, which operates in the high-latitude Northern Hemisphere at the 41 and

22 kyr periods to alter the pace of ablation of the ice sheets in the Northern Hemisphere. Some researchers have offered a significant criticism of the paradigm, despite the fact that almost everyone agrees on the significance of orbital forcing on the geological records of the previous 2 Ma. The fraction of the variance attributable to orbital changes never exceeds 20%.

There are currently no regional glacial chronologies with time-scales that can be easily compared to the global ice volume record preserved in the benthic isotope data. Since there is no glacial stratigraphy on land that preserves all of the 22, 41, and 100 kyr cycles that are inferred from marine records, the history of the continental ice sheets is largely inferred from the marine records. As a result, the glaciology of the climate response is unclear for specific ice sheets and glaciers. It would seem that modeling ice-sheet variations caused by orbital changes is the most productive approach to comprehending the interactions between the ice sheet and the climate. The rate of growth of large ice sheets is one of the major issues in the interactions between ice sheets and climate. The size of ice sheets is very sensitive to the distribution of mass balance and has generally been supported by earlier researchers. According to the Sawtooth model proposed by Broecker & Van Donk, the global ice volume gradually increased starting from MIS 5d at 115 cal. kyr and peaked in MIS 2 at about 20 cal.kyr (Crucifix, 2018). On the other hand, research on coral reef layers suggests that sea level dropped quickly in MIS 5d at the beginning of the latest glacial cycle. Recent research in Barbados indicates a fall of ≥37 m or 3 to 4 m kyr−1. This is roughly equivalent to constructing a modern Greenland Ice Sheet, or at least half of one, every 1,000 years, or almost half the volume of the LIS during MIS 2. A closer look at the temperature-precipitation connections in the GISP2 data reveals the inconsistent relationship between growth rate and climate. Colder environments are thought to equate with lesser accumulation. An early attempt at three-dimensional glaciological modeling on the instantaneous growth of the LIS succeeded in accounting for a 20-meter drop in sea level in 10,000 model-years by pushing mass balance to a limit approximately three times the present. These early model patterns of ice-sheet formation and expansion are comparable to those of modern glaciological models; however, they fail to accurately represent the rate of ice build-up.

What atmospheric and oceanic conditions are necessary to cause such a quick and positive increase in the net mass balance of the developing LIS is the crucial question. The growth of the LIS may be related to less sea ice in the Labrador Sea and Baffin Bay during glacial inception, given the strong positive correlation between the duration of the snowfall along the coast of Baffin Island. The existence of warm water mollusks in glacial marine sediments linked to the start of the last glacial period has served as support for some of these claims. Arguments like these have suggested that because of increased winter precipitation, global warming may lead Arctic Canada to glacierized again. The discovery that the Antarctic Ice Sheet was thought to be a Pleistocene in the 1960s serves as an example of the progress made in deciphering ice-sheet evolution. Large amounts of ice are thought to have accumulated in Antarctica as far back as 34 million years ago, around the end of the Oligocene. Over the next 14 million years, there was a period of increasing global cooling, which ended in the development of a fully formed Antarctic Ice Sheet in the middle Miocene, approximately 15 million years ago. The maximum ice sheet pushed out to the outer continental shelf before retreating to its current location, about along the coast of East Antarctica, 13.6 million years ago. According to some researchers, there has been a dynamic history of ice sheet expansion interrupted by periods of major deglaciation. Some experts have maintained that the ice sheet has been stable for more than 13 million years, with only coastal glaciers and ice shelves reacting to variations in global sea levels. The latter perspective is in line with the theory that the ice sheet has changed the local climate and ocean circulation so much that it has brought about a new level of stability in the system. Since global temperatures are still well below the melting point, warmer air does not greatly accelerate melting; instead, it increases snowfall and thickens glaciers. Since a large portion of the West Antarctic Ice Sheet was below sea level and therefore vulnerable to variations in the global sea level, the situation there may be different.

4.12 GLACIAL RETREAT

There has been progressive retreating over the past 10,000 years, apparently in response to sea-level rise following the loss of Northern Hemisphere ice sheets at the end of the last glaciation, even though there is no concrete evidence that it has disappeared since it too formed in the early Miocene. The shape and behavior of the ice sheets in the Northern Hemisphere have been the subject of intense debate despite a long history of research dating back to the mid-19th century. The Laurentide and Greenland ice sheets appear to have been vast, covering North America to a similar extent as modern-day Antarctica. Along with covering the Barents Sea and expanding as ice shelves into the Arctic, the Scandinavian Ice Sheet was also very large. However, the absence of snow in eastern Siberia limited the amount of ice, and glaciers were centered on mountain ridges. Overall, the accumulation of ice, most of which was in the Northern Hemisphere, was equal to a 120-meter drop in sea level worldwide. Although mountain glaciers have existed in the Northern Hemisphere for millions of years, they are primarily associated with the last 500,000 years or so. These cycles of huge ice-sheet growth and disintegration occur every 100,000 years(Denton, 2000). Recognizing the tremendous variability of the climate on time-scales less than ice-sheet growth and degradation has been of focus for the past few decades. The ice-core records document sudden temperature fluctuations, particularly Dansgaard-Oeschger episodes, which are characterized by a rapid warming of up to 10°C followed by a slower phase of cooling. These differences have highlighted potential changing mechanisms and the significance of compiling global records of glacier fluctuations.

CHAPTER 5

GLACIERS OF THE PAST

Contents

5.1 INTRODUCTION

Ice ages on Earth were thought to have been rather intermittent, with only one significant glacial episode (Huronian Glaciation Event) occurring at roughly 2.3 Ga, according to evidence found in sedimentary strata(G. M. Young, 2019). There are various possible interpretations for the limited early glacial deposits. Perhaps a high concentration of greenhouse gases in the atmosphere is what caused the very high surface temperatures. On the other hand, the lack of evidence in the geological record may simply reflect the lack of continental lithosphere, which serves as a foundation for glaciers, and the water-covered area may have simply frozen if the Archaean Earth had cold periods. Throughout the previous 750 million years, there have been four main and several minor glacial events, making glaciations far more frequent. When one considers that the sun was likely somewhat less luminous/bright in the Archaean era, the lack of Archaean glacier deposits becomes even more puzzling, prompting many to speculate that the early Earth's atmosphere was with greenhouse gasses like carbon dioxide. The largest glaciers are the Cryogenian, Permo-Carboniferous, and Paleogene to recent. If it weren't for the fact that humans originated during an ice age, we could have completely missed the existence of ancient glaciers. Only because of the zeal and tenacity of pioneers like Agassiz were topographical and sedimentological standards established for the identification of the previously vast ice sheets in Europe and North America. Since it was widely believed that the continents had just been covered by km-thick ice sheets, there have been numerous attempts to distinguish between comparable deposits found in sedimentary rocks and material left by Pleistocene glaciers. The discovery of conglomerates in the Dalradian (Precambrian) sequence on the island of Islay, in western Scotland, by Thomson is generally regarded as the first report of glacial deposits in Precambrian rocks. Thomson believed the conglomerates to be evidence of ancient glacial activity. Dwyka Formation and its corresponding glacial deposits from the major Permo-Carboniferous event were important pieces of evidence that some researchers utilized to strengthen the then-controversial hypothesis of continental drift.

5.2 GLACIAL DEPOSITS

Glacial deposits in Precambrian rocks can be identified using simple field techniques. Among these are huge diamictites, which are conglomerates of different clast sizes and types that are floating. Rock flour gets its name from the crushing of rock material that occurs when abrasion occurs in and at the base of a moving ice mass. Similar to striated substrates, when clast-bearing ice masses dominated rocks or sediments, striated and faceted stones are likewise used to identify glacial action. Many other characteristics have also been used to identify terrestrial glacial activity; however, because they are easily eroded and frequently fold and metamorphosed, they are hardly preserved in Precambrian glacial deposits. It is believed that most old glacial deposits originated in oceanic environments. The Chemical Index of Alteration or CIA is one important geochemistry component that has been extensively utilized to determine the character of materials that are often composed of old glacial deposits. Age data from detrital zircons in glaciogenic rocks may be utilized as a marker in some conditions. Lower surface temperatures have been attributed to both extraterrestrial and terrestrial sources. Variable solar radiation, Saturn-like rings forming a disc-shaped body around the Earth that casts shadows on some parts of the planet, the solar system passing through the galaxy's spiral arms, and encounters with nebulae, such as supernova fragments/remnants and dark clouds are a few examples of suspected extraterrestrial causes. Since some researchers proposed that the Sun's radiative power has been increasing throughout geological time by drawing comparisons with the known life cycle of other stars, it has been widely accepted that the lack of glacial deposits in Archaean rocks is due to a high concentration of GHGs like CO_2 or methane in the early atmosphere (Nisbet & Fowler, 2011). The occurrence of a CO_2-rich atmosphere on the early Earth is supported by geological evidence such as the formation of deep palaeosols and extensively weathered sedimentary rocks like quartz arenites in the Archaean and Palaeoproterozoic periods.

5.3 RESEARCH FINDINGS

Supercontinents are prone to weathering due to their thermal buoyancy, which is significantly increased during some periods linked to supercontinent breakdown, as well as during orogeny related to assembly. The Huronian and Cryogenian ice ages occurred during the disruption or division of glaciations concerning the supercontinental cycle. In contrast, the great Phanerozoic glaciations (Permo-Carboniferous and Paleogene-Neogene) were caused by orogenic activities linked to collisions during supercontinent assembly. The causes of this are not immediately clear, but they might indicate a greater vulnerability to glaciation as the atmospheric CO2 level decreased over time, perhaps causing younger glaciations to form sooner in the supercontinental cycle. The majority of Precambrian samples accumulated at low paleolatitudes, and many formed in marine locations, indicating that glaciers were able to exist at, or close to, sea level in tropical regions, according to data collected over the last few decades concerning paleolatitudes at which ancient glacial sediments were deposited. Although glaciers are known to exist in low latitudes, they are unable to withstand the high ambient temperatures there. As of right now, such conditions are only known from polar latitudes. Comparable research on

Phanerozoic glacial deposits has produced different findings, indicating that they originated in high latitudes similar to those found in Antarctica today.

The high obliquity theory (HOT) states that due to the drastically altered annual distribution of solar energy at the surface, glaciation would have first affected areas at low latitudes if the Earth entered a period of low surface temperatures. There is still debate over the issues raised in the case of the glacial deposits of the Proterozoic at low paleolatitudes. The lack of glacial evidence at high latitudes is still a problem, despite the fact that it has been linked to a shortage of continental lithosphere in the circumpolar areas. Freezing of the polar oceans would have been the cause of cold conditions. It is assumed that the Earth reached a 'runaway albedo' condition as ice moved toward lower latitudes, resulting in surface and massive ice sheets covering the continental masses, even though they were mostly found close to the equator. Precambrian basins would have been deposited as a result of melting the ice sheets, as this sequence of events is unlikely to have occurred due to the requirement of a massive source of water for the build-up of thick, extensive ice sheets. Cryogenian (Sturtian) paleogeography is used to estimate climate conditions (Yu et al., 2020). Growing glaciers at low latitudes would have been fed by the polar ocean as the Earth entered an ice age, leading to their expansion into higher latitudes. The end of these massive Precambrian glaciations to the buildup of atmospheric CO_2 from ongoing volcanic activity and decreasing drawdown due to less weathering and photosynthetic activity. Nonetheless, the Gowganda Formation and the earlier glaciogenic units differ greatly from one another. The earlier diamictites contain a quartz-rich matrix, and the stratigraphy of the lower diamictite-bearing formations is simpler than that of the thicker Gowganda Formation. The extensive Gowganda and its formations provide clear evidence of marine conditions. It was feasible to distinguish between marine deposits to the south and continental crusts to the north. The Gowganda Formation in its northern region (outcrop belt) is marine. Researchers provided a thorough description of the stratigraphy and sedimentology of the southern Gowganda Formation. This investigation revealed the intricate stratigraphy of the formation. It encompasses two glacial advance-retreat cycles, divided by a dense, finely bedded, nonglacial argillite layer.

Except for the Hurwitz Group on the west side of Hudson Bay, all of the Palaeoproterozoic rocks may have formed along the continental margin and appear to have been deposited during the breakup of a supercontinent. The Huronian Glacial Event is classified as a B-type glacial since it occurred during the breakup of a supercontinent, at least in North America. Palaeoproterozoic glaciogenic deposits were found in Western Australia. Mass flow deposits and turbidites are also present in the Meteorite Bore Member of the Kungarra Formation. The iron-rich rocks of the Hamersley Group may also have glaciogenic deposits and we need to consider these deposits to be evidence of an earlier glacial episode. The Meteorite Bore Member is between 2.45 and 2.21 Ga old. Two distinct glacial units within the Kungarra Formation of the Turee Creek Group are supported by recent investigations of Western Australian zones. Although the Paleoproterozoic glacial deposits in Western Australia and South Africa have an unclear basin setting, some researchers suggested that this region was rather stable, suggesting that the incidental fluctuations in sea level were likely glacio-eustatic. It isinteresting to contrast these findings with the Gowganda Formation, which exhibits evidence of two distinct glacial episodes separated by a thick layer of argillite that displays no physical signs of glacial influence. Thus, the stratigraphy of the Gowganda Formation can be understood in terms of climate and sea level variations comparable to those in Western Australia, although the formation was deposited in a high tectonically active environment. The exact location of glacial deposits in Western Australia and South Africa is still unknown.

5.4 GLACIATIONS

Most likely the most extensive glaciations in history occurred during the Cryogenian period. Although the idea of extensive Precambrian glaciations is considerably older, these glaciations are ascribed with the name 'Snowball Earth'. There is stratigraphical and geochronological evidence of at least two massive glacial episodes, known as the Sturtian and Marinoan glacials from South Australia, in many parts of the world. Researchers estimated that the Cryogenian events were between 727 and 654 Ma and between 660 and 635 Ma, allowing for a minor temporal overlap between the two glaciations (Chumakov, 2009). Many Sturtian glacial deposits accumulated in rift basins that indicated the break-up of the supercontinent Rodinia. Iron deposits are connected with several Sturtian glaciogenic deposits, which is noteworthy because these rocks are absent from the geological record for the preceding billion years. The deposits of ferrous iron in oceans covered the oxygenated environment by a dense layer of ice is what caused these iron formations. Restoring oxygen to the oceans once the ice melted resulted in the precipitation of iron. The Sturtian iron formations originated as a result of hydrothermal activity in rift basins formed during the disintegration of Rodinia. The association of carbonate rocks, primarily dolostones, on diamictites is another peculiar connection that is most frequently observed in the Marinoan glacial deposits. Some researchers have connected the unusual conditions toward the end of the Marinoan glacier, when atmospheric CO_2 levels are considered to have been high, to the carbonate rocks (R. He et al., 2021). Following the proposal of low depositional palaeolatitudes for Cryogenian glacier rocks in South Australia, several comparable findings from Neoproterozoic glacial deposits around the world have been documented. One of the primary views of the SEH (Snowball Earth hypothesis) was the deposition of glacial deposits near sea level in low latitudes, however, some researchers provided an alternative explanation. They proposed that the presence of ice sheets in tropical latitudes was caused by a severely altered distribution of climatic zones, which was brought about by a rise in the Earth's axis. It was believed that the annual heat distribution throughout the surface would lead glaciers to form preferentially in tropical latitudes if the Earth entered a glacial phase when the tilt of its axis (obliquity of the ecliptic) exceeded 54 degrees. Although there seems to be a difference between the palaeolatitudes at which the Huronian and Cryogenian glaciations occurred and those of the late Ediacaran and all Phanerozoic, it is uncertain by what mechanism the Earth reached its current obliquity. The two great Cryogenian glaciations seem to have occurred during the disintegration of a supercontinent, under tectonic conditions akin to those of the early Palaeoproterozoic Huronian Supergroup. There is evidence supporting glaciation in the form of deposits from the Early Paleozoic era. The majority of the early Paleozoic record is limited to the Ordovician, however, certain locations have evidence of early Silurian glaciations. A significant portion of the sedimentary record is preserved in what is now North Africa and Arabia, offering an interesting distinction between the arid environments of today and the ancient, 443-million-year-old (Hirnantian) glacial record (Le Heron et al., 2018). Due to the dispersion of outcrops in these desert environments, rock exposure is usually very good, allowing for the documentation of old glacial landforms. The glacier record from the Hirnantian is remarkably sandy when compared to other Precambrian and Phanerozoic glacial records.

5.5 GONDWANA

The development of Gondwana caused considerable weathering, intracratonic mountain building, and broad terrane amalgamation. One of the main causes of the Hirnantian glaciation is thought to have been the decrease of atmospheric CO_2 brought on by increased biological activity and weathering. The Hirnantian glacial period may have begun as a result of a fall in atmospheric CO_2 caused by an increase in weathering activity. Furthermore, the development of elevated terrain may have expedited the production of alpine glaciers, contributing to an elevated rate of weathering. It is thought that the opening and closing of oceanic pathways plays a significant role in cooling and may have been a major factor in the Hirnantian glacial. Gondwana may have been shielded from the warming effects of circumpolar currents by an isthmus in central America, which would have stimulated insulation and cooling. The Late Ordovician glacial deposits in North Africa are named differently across regional boundaries and they exhibit internal complexity. The Libyan successions surrounding the borders of the Murzuq Basin in the west and Al Kufrah Basin in the east have received attention in research (Aziz, 2000). The NW, Central, and SW parts of Saudi Arabia are home to early Paleozoic glacial deposits. The syn-glacial sediments in Southeast Turkey seem to have an ice-margin character. The predominant pattern in basal deposits is a fining upward and is transitional. These glacial deposits begin with channelized sandstone that is deposited in ice-proximal settings and terminate with sandstone, siltstone, and shale with a few lonestones that suggest glaciomarine situations with uncommon dropstones. Conversely, some researchers characterized the glacial deposits in Iran as stratified diamictite units with a high concentration of clasts, interspersed with sandstone that is fine to coarse-grained and moderately sorted. Many areas of modern-day Europe were terranes or maybe joined to the Gondwana landmass, placing them in a peri-Gondwanan toward the modern-day north of Africa. Thus, Early Paleozoic glacial deposits are found throughout southwest Europe, including the Armorican Massif, the Iberian Peninsula, Thuringia, and the Prague Basin in the Bohemian Massif. According to several studies, they are named glaciomarine deposits of the Hirnantian age. The materials consist of diamictites and shales that have been dumped in a shallow marine shelf, containing large clasts known as dropstones.

South America has glacial deposits from the early Paleozoic era (Limarino & Spalletti, 2006). Based on detrital zircons, they are found in central and northwest Argentina. They can be found as glaciomarine facies in the Peru-Bolivian Basin and as continental deposits in Brazil. Continental deposits can be found in Brazil's Parnaiba and Parana basins. Based on shelly fossils found in the layers above the glacial deposits, researchers inferred the age of the oldest Silurian glacier in Brazil. Younger glacial events in the Amazonas Basin were dated as early Telychian and late Telychian to earliest Wenlock. The Gondwana Super-fan System, a layered siliciclastic succession of rocks, is the substrate primarily covered by the ice sheets that expanded across the southern supercontinent of Gondwana. Super-fan deposits are found in North Africa and Arabia, and they are characterized by very similar stratigraphy in Cambrian through Ordovician strata. In this period, which stretches from Morocco to Turkey, intercalated shallow marine and fluvial controls predominate. A thorough analysis of each outcrop belt in between reveals a similar pattern. For example, all studies of glacial deposits in southern Libya indicate that they rest unevenly on shallow marine sandstones from the mid-Ordovician to early Hirnantian and mudrocks. The homogeneity of the substrate over which the Hirnantian ice sheets progressed is significant on a regional scale. This homogeneity had an immediate effect on the sedimentary record. It is possible to claim that the stratigraphic architecture and composition of the substrate along Gondwana edges had a significant impact on the dynamics of the palaeo-ice sheet as well as the composition of the strata formed during the Hirnantian glacial. Scientists proposed that at least five glacial cycles have been identified in the North African region.

5.6 GLACIAL MAXIMUM

Some researchers suggested a map of the Hirnantian Glacial Maximum (HGM) that used satellite imagery, subsurface data (seismic observations), and observations from outcrops throughout North Africa. Paleo-ice stream paths were hypothesized to exist. Very high-quality mapping interpretations from desert regions of North Africa, including the Murzuq Basin in Libya, have validated some of these speculative palaeo-ice stream paths. The Sarah Formation is a high-quality glacial deposit archive found in the northwest of Saudi Arabia. The glacial characteristics include distinct instances of intraformational striated surfaces, which are supported by marks in certain locations, indicating that the structures originated in soft sediment. The ice age began earlier than the Quaternary. The Tertiary Period had a long-term trend of cooling global climates. This era is marked by the opening of the North Atlantic Ocean and the concurrent closing of the Tethys Ocean, which is aligned east-west. Significant palaeogeographical rearrangements and significant modifications to the circulation patterns of the ocean-atmosphere system were brought about by this evolution. The distribution of climate is specifically caused by the drift of Antarctica to its position over the South Pole and the isolation of the Arctic Ocean. During this period, the uplift of young mountain belts, such as the Alpine-Himalayan chain, and the drift of the continents limited the amount of energy that could be transferred to the polar regions. Glaciation was one of the effects of the distribution of continents and oceans on the surface of the Earth. Other than sporadic activity in the Eocene in both the Southern and Northern Hemispheres, major glaciation started in eastern Antarctica in the late Oligocene. The Miocene witnessed mountain glaciation in Alaska, Greenland, Iceland, and Patagonia; the Pliocene witnessed it in the Alps, the Bolivian Andes, and probably Tasmania. Early Pleistocene, the beginning of the Quaternary, was marked by climatic oscillations dominated by the 41-ka orbital precession cycle, in which comparatively few cold periods were long and cold enough to support the formation of substantial ice sheets. Currently, only 14 of the 41 cold phases from that era exhibit signs of significant glaciation. In both the North Atlantic and North Pacific oceans, extensive ice rafting is present from the oldest cold phase, indicating that glaciers have reached sea level (Hebbeln et al., 1998). The onset of synchronized glaciation in the regions of Greenland, Scandinavia, and North America dates back to 2.72 Ma. The Early-Middle Pleistocene transition or EMPT, started around 1.2 Ma and took about 800 ka to fully establish. It was during this transition from dominant orbital cyclicity to the 100 ka cycles that the cold periods, or glacials, became consistently cold and extended enough to support the development of ice sheets outside of the polar regions on a continental scale. It is possible that within the last 2.6 million years, there were at least 20 periods in which extensive glaciation may have formed, with the most extensive occurring just in the last 900 ka. Less extensive glaciations tend to be removed by subsequent, more extensive ice advances, pinpointing the exact location and extent of these glaciations is exceedingly difficult. The preservation capability of older sequences rapidly decreases with time and subsequent glaciers, even if this is true over all terrestrial areas. This is particularly true in mountain locations. Based on an analysis of glacial frequency across the Cenozoic, it can be inferred that the establishment of glaciation in the Southern Hemisphere began in Antarctica and South America, where it persisted continuously from the early Neogene to the present. In contrast, glaciation in the Northern Hemisphere grew significantly at the start of the Quaternary, increased again in frequency in the last stage of the Early Pleistocene, and reached extremely high levels in the Middle-Late Pleistocene. The remarkable increase in ice-sheet extent during the Quaternary shows that global glaciation is essentially a northern hemisphere phenomenon based on the distribution of land mass in higher latitudes.

5.7 NORTH AMERICA AND OTHER COUNTRIES

There is ample evidence of glaciation from the Quaternary and even the Neogene in the Northern Hemisphere. Only Alaska and the neighboring Canadian North-West territories, together with Greenland and the Rocky Mountains, have the longest stratigraphic sequences, which show evidence of glaciation from the Neogene to the present. Regionally, glaciation occurred in the Pliocene and periodically during the Pleistocene, with the accompanying IRD in marine areas found in northern Canada and Alaska dating from the Early Miocene. A similar sequence is observed in neighboring British Columbia, especially in the north. Similar to this, glaciation started in the Miocene in Greenland and Iceland, continued regularly during the Pliocene, and continues to this day in the mountain regions. Similarly, glaciation is known to have existed in Norway during the Early Miocene, Early Pliocene, and Plio-Pleistocene, as well as in its nearby regions and the Barents Sea. IRD from a core from the Norwegian-Greenland Sea off East Greenland dates from approximately 38 to 30 Ma, at least some of which is most likely from glaciers rather than sea ice. Svalbard is excluded by certain features of this debris, which indicate an East Greenland Current (EGC)(Katrine Elnegaard Hansen et al., 2022). This IRD could be isolated mountain glaciers or a larger area of covered ice sheets. While adjoining ice sheets have existed since the early Pleistocene, inland ice sheets were intermittently found in Greenland by the Late Miocene, particularly in the east, with ice reaching the sea in SE Greenland. There is a considerably shorter glacial sequence in the eastern Rockies of the United States. Here, Plio-Pleistocene-aged fossils from North Dakota, California, and Montana are known. The first glacier on Mount Kilimanjaro in East Africa is estimated to have existed around 2.0 million years ago. Before the Middle Pleistocene, glaciers only left ice-rafted remnants in parts of Europe that are not mountainous. Volatile minerals found in the Netherlands suggest that the Baltic region experienced substantial glaciation towards the end of the Early Pleistocene. In the Alps, glaciation has also been established since the Plio-Pleistocene.

5.8 SUBGLACIAL PROCESSES

Ocean basins and glaciated continents are dominated in terms of area and thickness by subglacial processes and sediments. Subglacial processes of deposition and erosion are intricate. Because of the inaccessible environment beneath ice masses, our understanding of the subglacial environment, its processes, and its landforms is incomplete. Subglacial deposits are extremely important. It is impossible to overstate the topographic and volumetric abundance of subglacial sediments in countries with glaciers. These sediments cover large portions of all continents where glaciation has occurred since the beginning of geological time. Subglacial sediments, their structural architecture, grain size, and stratigraphic complexity are all important for the foundations of roads, railroads, airport runways, sewage disposal sites, toxic waste sites, agricultural land use, groundwater distribution and movement in aquifers, and the use of glacial sediments in drift mining. Furthermore, the use of glacially derived construction materials in areas last glaciated in the Pleistocene is dependent on subglacial sediments. The demand for applied knowledge of glacial sediments and environments will increase in the twenty-first century as our awareness of environmental hazards and rational land use grows. Knowledge of the subglacial environment will be of utmost importance. Large population centers have historically grown around glaciated regions due to the fertile soils that encouraged agriculture and provided food for huge numbers of people. The subglacial environment possibly has the most impact on the lives of millions of people in any glacial ecosystem. The subglacial environment is thought to be made up of several sediment layers with boundary interfaces between an upper ice mass and an underlying bed. These layers are then surrounded by an intricate process that interacts to change the morphological, thermal, and rheological conditions of the sandwiched sediments, sometimes leading to the evolution of interface junctions. The processes of glacial erosion and deposition are governed by the nature of this contact. Understanding the dynamics of this intricate network of contact points is essential to understanding subglacial bed types and related formations. With each advance and retreat of the ice masses, these boundary zones move across the terrain.

An understanding of the sediment delivery systems (SDS) in supraglacial, proglacial, and subglacial environments is necessary (Menzies et al., 2018). The final outcomes of these sediment systems are landforms and glacial deposits. Sediment delivery is dependent on many factors, including ice dynamics, the kind of ice mass, base and subjacent geology and sedimentology, the temporal and spatial variability of sediment discharge (flux), related hydrological regimes, and topography. The ice interior and basal thermal conditions, regardless of whether they are polar or temperate, determine the specific local sediment delivery systems. Sediment transfer under polar frozen bed conditions was formerly believed to occur primarily in the supraglacial system. But current research indicates that limited transit does occur as part of a slowly deforming bed even under polar or subpolar bed conditions. Large amounts of sediment are transported in all sediment subsystems in temperate bed conditions as opposed to frozen bed conditions. There are variations in the amount and pace of sediment transport within an ice mass due to diurnal, seasonal, and annual cycles. Transport of suspended sediment and bedload within the supraglacial, englacial, and proglacial environments increases daily as meltwater flow rises in response to solar radiation. Local weather patterns affect seasonal variations in sediment transport. Meltwater transport ceases in the supraglacial and englacial regions when below-freezing temperatures prevail. Simultaneously, subglacial transport is anticipated to be limited, with minimal or no extra surface meltwater; however, in temperate conditions, frictional heating and geothermal effects sustain a certain degree of subglacial meltwater transport. There is still much to learn about the complex sediment transport channels found within glacier and ice sheet systems. Subglacial environments experience substantial temporal and geographical fluctuations in sediment transport processes and flux rates. Sediment flux rates below current ice sheets, ice streams, and ice shelves have not received as much attention as flux rates in valley glacier systems. Considerable sediment transfer by ice sheets is caused by swift-moving ice streams. But these are relatively small, fast-moving zones of ice, so it shouldn't be assumed that basal ice elsewhere doesn't move a lot of sediment. Estimating such sediment flux rates requires attention because there is currently very little evidence of subglacial conditions available.

5.9 ROCKS AND GLACIERS

Rarely do isolated glacial striae influence entire rock surfaces. Striated rock surfaces and clasts carry multiple striations, frequently of varying directions. The pressure on one side of a bedrock that meets basal ice will cause pressure melting, which will allow water to enter rock crevices on the other side of the rock and refreeze. As a result, a block or a portion of the bedrock would get frozen and be cut. Refreezing, however, is far less likely to happen under the entire glacier or on a large scale under temperate ice. Adfreezing occurs at the top of the sediment since many glaciers are triggered by moving basal debris, which is occasionally linked to the basal ice and in other spots separated from the basal ice (Waller et al., 2012). The large structures created by glacial erosion are fjords and basins. Fjords are exclusively found in rocks; however large basins can occur in both sediment and rock. They are restricted to hilly regions that were once covered by ice sheets. Not only does glacial erosion occur on bedrock, but it also erodes sediments. Erosion can happen at the ice/bed contact, in the sediment, and at the base of a deforming bed. It can be claimed that eroding glaciers can practically move mountains. Since cold-based ice was believed to be frozen to the bed and to have movement occurring within it, it was previously believed that cold-based ice had no erosional effect. As evidenced by recent research conducted in the Antarctic; cold-based ice does erode its bed. Supraglacial debris can become englacial or even basal debris when it enters the glacier through crevasses or is washed in by supraglacial meltwater. Subglacial debris can travel or melt the top to become supraglacial. It transforms into basal debris by transit in meltwater or basal melting. Similar to englacial debris, supraglacial debris can also form from basal debris. It goes without saying that all of the debris is mixed when a glacier falls into an icefall. Even though supraglacial debris can usually be easily identified from basal debris, each level may nevertheless contain material that is indicative of the other due to the mixing process. Rounded clasts and boulders that have been transported and deposited from supraglacial environments dependent on bedrock sources are frequently found in boulders and ablation deposits in many glaciated areas. This process is known as debris entrainment, and it is caused by the removal of rock and debris, which causes clasts to rise within the ice to the supraglacial.

5.10 DEBRIS AND MECHANISMS

Debris that has joined the glacier at different places, either on the surface or from the sides and from below, will not reach the frontal margin at the same time due to the typical velocity profile of a glacier or ice sheet. Particles of debris, particularly those at the base, will come into contact with one another because of the differential velocity. Particles may erode one another depending on differences in hardness and whether the ice that holds them would melt under pressure. Larger particles as a result become striated and polished, occasionally breaking up. Many tills are thought to be sandy as a result of these comminution processes. Mass transport of sediment at the bed must be substantial since forward movement in the bed might contribute significantly to the overall glacier velocity. The bond between debris within the ice being released through melt-out, the effect of meltwater melting the ice, or the immobilization of subglacial debris at the basal interface between the bed and active ice is the basis for the mechanics of subglacial deposition of debris from ice masses. This latter process may be the largest in terms of volume. Gravitational debris flows released subglacial or the immobilization of preexisting subglacial silt under soft deforming bed conditions are two key additional sources of sediment deposition. Sediment release/flux primarily occurs when debris presses on the bed or other debris and its internal frictional strength is larger than the force exerted by the active ice or mobile sediment moving forward. Debris can either become lodged or immobilized as a result of decreased external stress levels or increased internal shear strength of the material. Additional release mechanisms take place when the ice stops moving forward, becomes passive, and melts out gradually through basal melting. Particles that have lodged or been ejected may reactivate or become encased in more debris, which would cause them to merge with the deforming substrate. Then, deformation will destroy most or all of the earlier structures, except for the largest boulders at the surface. The distribution and structure of the debris in the ice should be preserved throughout the melt-out. Melt-out structures will probably be completely destroyed or severely damaged if the deposit is reactivated in the deforming bed by new glacial activity. Since melt-out deposits are usually vulnerable to subsequent changes within dynamic glacial environments, they may have a limited survival potential in places with low glacier bed relief.

Subglacial sediments are classified as unsorted (tills/diamicts/diamictons) or sorted (glaciofluvial/ glaciolacustrine). The latter can occasionally be challenging to distinguish in terms of type, origin, and location. Since tills are polygenic and generally considered to be of continental (on land) origin, they can be mistaken for glaciomarine/glaciolacustrine diamictic sediments in some places. In other places, however, they may be associated with volcanic lahars, intercalated debris flows, and both land-based and marine flow tills. Because they are diamicts, locally generated materials often make up the majority of the subglacial till matrix (D. J. A. Evans et al., 2016). However, in areas where a fine-grained matrix is created, these clay- and silt-sized sediments may persist within sediments for hundreds of kilometers. The sediments are generally subangular to subrounded, with striae on their surface facets, and a broad variety of grain sizes, from coarse sands to clays, clasts (pebbles) to small boulders. Large block units of sands and gravels, subglacial tills are stratigraphically composed of intraclastic units that, like many of the larger clasts and boulders, float within the sediment matrix. Many intraclasts can be seen at different angles while others have a boudinage appearance. These tills had few or no big clasts. The matrix color varies in many locations, which can be attributed to sediments with varying sources. Tills are frequently categorized as massive at the macroscale, but upon closer inspection, they show structure, are complex in strata, and display bedding, lines, and a variety of complex microstructures at the microscale. Tills can originate from active temperate wet-based and polar cold-based ice masses, under stagnant ice conditions, and subaquatic environments of both attached and floating ice masses. They can form within subglacial, englacial, proglacial, and supraglacial glacial subenvironments.

CHAPTER 6

GLACIER MECHANICS

Contents

6.1 INTRODUCTION

Glaciers can be categorized using their shapes. Valley glaciers are long, somewhat narrow glaciers that flow essentially in a single direction i.e., downstream. A valley glacier is referred to as a tidewater glacier as it approaches the coast and interacts with the sea.

Figure 6.1 Cirque glaciers

Source: https://en.wikipedia.org/wiki/Cirque#/media/File:Circosabisko.jpg, Creative Commons License

Cirque glaciers (Figure 6.1) are very short valley glaciers that may only cover a small basin in the mountains (Evans, 1977). Glaciers are referred to as ice sheets if they are sufficiently huge ice caps. Some ice caps on Canadian Arctic islands, for instance, provide feed for outlet glaciers, which are essentially valley glaciers moving away from an ice sheet or ice cap. Valley glaciers and ice sheets usually differ in other important ways. Glaciers are also categorized based on their thermal properties. It is common knowledge that water freezes at 0°C, but we sometimes forget that once all the water in an area has frozen, the resultant ice can drop below that temperature as long as heat can be extracted from it. As a result, in cold areas, the temperature of ice in glaciers can be much below 0 °C. These glaciers are known as polar glaciers. More precisely, polar glaciers are those where, with the possible exception of the bed, the temperature is consistently below the melting point of ice. It will be appropriate to refer to such glaciers as Type II polar glaciers and to those that are frozen to their beds as Type I polar glaciers.

6.2 POLYTHERMAL GLACIERS

Polythermal or temperate glaciers are those that are not polar (Blatter & Greve, 2015). Subpolar glaciers, also known as polythermal glaciers, have significant amounts of cold ice mixed with substantial amounts of ice at melting temperature. The cold ice is typically found on the lowest portion of the glacier as a thick layer. The reason glaciers occur is that there are places, usually in polar latitudes, where the amount of snowfall in the winter surpasses the amount of meltwater and other losses in the summer. The accumulation region is the term given to this portion of the glacier as a result of the net accumulation that occurs. Snow transformation is the outcome of molecules moving between the liquid and vapor phases. Firn is the term for snow that has been affected by these processes and is older than a year. Solid ice is the final product of the firnification process, which often takes several years. Gravitational forces push this ice toward locations with lower altitudes where it can move. Eventually, this causes the ice to accumulate in areas where annual melting exceeds snowfall. In this case, summertime melts all of the winter snow and a portion of the underlying ice. We refer to this as the ablation area. The equilibrium line is the boundary that separates the areas of accumulation and ablation at the end of a melt season. Melt during the most recent summer precisely matches net snow accumulation during the preceding winter along the equilibrium line. Snowflakes begin to turn into ice when water molecules freeze. This process causes the snowflakes to become spherical, or rounded, which lowers their surface area and, in turn, their free energy. The rate at which this rounding occurs increases with temperature. Sintering, which is a step in the process of further compaction, includes the diffusion of materials within the grains, the nucleation and development of new grains, and the internal deformation of the grains. When pore spaces are still quite large in the transformation process, sublimation plays a significant role. The importance of internal deformation grows with increasing pressure and depth of snow cover. The densification process accelerates in warm climates because of the possibility of surface tension drawing grains together when water films form around them and the possibility of percolating melt water filling air voids and refreezing.

6.3 THE IMPACT OF THE SEASONS

At a density of about 830 kgm^{-3}, a significant change in the transformation process takes place. The pores close at this density, stopping more air from passing through the ice. It is possible to approximate the elevation of the ice surface by knowing the depth in the glacier at which this occurs. Depending on the temperature, pore closure can occur at depths of a hundred meters. There are places on high-altitude polar glaciers, like the Antarctic or Greenland ice sheets, where summertime melting does not occur. There is some melting at relatively lower elevations, and the resulting meltwater percolates downward into the cold snow where it refreezes. Parts of the snowpack on an ice sheet with distinguishing characteristics are called facies (the dry-snow facies and the percolation facie) (Wolken et al., 2009). The dry-snow line, which separates these facies, is roughly located at the elevation where the mean temperature is -6°C. Summer melting is enough to soak the entire snowpack at lower elevations. This snow forms a solid, porous coating upon refreezing. The base layers of the snowpack in this zone may get soaked with water. This water-saturated snow may refreeze and produce overlaid ice if the underlying ice is cold. Superimposed ice can be easily identified as long as it is still undamaged. The term superimposed ice zone refers to the area at even lower elevations where the melt season ends and only superimposed ice remains. The equilibrium line is the lowest limit of the superimposed ice zone at the end of the melt season. Similar to polar ice sheets, the water that seeps into cold snow at the start of the melt season on ordinary alpine glaciers may refreeze to form lenses and ice slabs. But the whole snow pack will have warmed to the melting point by the end of the melt season. As a result, these glaciers lack both the percolation and dry-snow facies. Furthermore, very minute superimposed ice forms on a temperate glacier because there is minimal heat transfer into the glacier beneath the snow cover. The release of latent heat throughout the refreezing of the water to infiltrate accounts for the majority of the warming of alpine snow packs. In contrast, heat conduction from the surface is negligible. A clear vertical zonation exists at any given location on a glacier in addition to the zonation in snow stratigraphy with altitude, especially on polar glaciers. The autumn snow in an annual layer has a higher vapor pressure than the winter snow because it is warmer. As a result, there is a differential in vapor pressure, which causes molecules from the autumn to diffuse into the winter snow. As a result, the autumn snow gets coarser and may become less dense. We refer to these layers of coarse fall snow as 'depth hoar'. Because depth hoar layers may be identified in snow pits and ice cores, they are significant. Glaciologists have used these stratigraphic markers to calculate accumulation rates in deep cores from these ice sheets, which have been measured over several years or decades in many parts of Antarctica and Greenland. Researchers found that the accumulation rate high on the Greenland Ice Sheet increased by the end of the Pleistocene and that the change occurred in a period of only three or four years. This is one of the most remarkable examples of the use of such physical stratigraphy. Due to the warming of the climate, there was an increase in the accumulation rate, and the retreat of the ice sheets was a result of this.

Extensive laboratory analyses of core samples can also be used to date ice. The method that is most frequently applied to this is the measurement of δ18O fluctuations. Winter snow has more negative δ18O values than summer snow because the air is colder and there is less of the heavier oxygen isotope, δ18O, in the snow. Consequently, a sinusoidal variation in δ18O was observed in a series of samples obtained from a single annual layer. When this method is applied to date very old ice, a huge number of samples need to be examined. Nonetheless, in ice older than 8000 years, annual layers have been found that are compressed but yet identifiable due to their isotopic changes (Mix et al., 2001). Seasons also affect the electrical conductivity and microparticle concentration of ice. Seasonal fluctuations in the aerosol composition occur. The summertime entrainment of dust by wind, when outwash plains and similar surfaces are clear of snow, is what causes the seasonality in microparticle concentration. We can use either of these versions for dating. Errors arise when some annual layers either contain two cycles of change in the parameter being used or do not vary at all when such techniques are employed to date relatively old ice. Nonetheless, layers of volcanic ash are commonly discovered in cores, and an exact date can be given to the ash-containing ice when its chemistry allows for the attribution of a layer to an eruption period. The precise age of thousands of years old ice has been determined. A glacier mass balance can be described in many ways depending on the condition. The amount of snow that falls throughout the winter months is known as the winter balance. On the other hand, the amount of snow and ice lost to melt is known as the summer balance, which is a negative number. The total of the winter and summer balances is the net balance throughout a year, which is typically understood to run from the end of one melt season to the end of the next. These balances are typically stated as the thickness of a layer of water.

6.4 WINTER AND SUMMER FLUXES

In the summer, significant amounts of accumulation can occur in the accumulation zones of polar glaciers, and in the winter, melt can occur in the ablation areas/zones of some temperate glaciers. On tropical glaciers, where deposit and melt can vary over periods of hours to days, this is most evident. Stakes inserted into specially drilled holes in the glacier to measure the height of the snow or ice surface are perhaps the most popular method. To determine the winter balance, measurements are taken at the end of the first melt season. To determine the summer and net balances, measurements are taken at the end of the subsequent melt season. To translate the winter accumulation and summer snow melt into water equivalents, measurements of snow density are also required. At the tips of valley glaciers that end in lakes and tidewater glaciers, cliffs form. Ice blocks of all sizes, from single crystals to hundreds of cubic meters, break off from these cliffs and drift off to melt in distant locations. This is called calving. The cliffs may reach depths of several hundred meters below the water surface, although they normally stand 60–80 meters above it. Certain outlet glaciers in Antarctica and Greenland have floating termini. Calving is an important, if not the primary mechanism of mass loss on most of the glaciers. For instance, calving from outlet glaciers that end in fjords is thought to be responsible for around 50% of Greenland ice loss. The size of the ice blocks that tidewater glaciers lose and the subaerial height of their faces are determined by the properties of the ice in their terminus. Since the ice is often temperate, water is present around the edges of the crystals, weakening the ice. Furthermore, the snouts usually have a lot of crevasses. The size of the ice masses that these glaciers release is constrained by some variables.

6.5 CALVING

The height of the calving face may be limited by the crushing force of a free column of temperate ice with substantial intergranular water, which can be attained at 60–80 meters. Disruptions are caused when ice pieces break off below the water surface and float upward to the surface. The amount of ice produced by these underwater events, however, may not be enough to explain the pace of retreat of the subaerial component of the terminus (observations of calving events on San Rafael Glacier in Chile) (Warren, 1993). This implies that a significant portion of the process known as calving may involve melting below the water surface. This is supported by the finding that the high calving rates occur in October when the water is hot. If a glacier is sliding over its bed at the pressure melting point, frictional heating from the sliding and deformation in the basal ice can melt huge amounts of ice. The specific net balance at an elevation of around 400 meters on the lower portion of Columbia Glacier is approximately $-4.5\ \mathrm{m\ a^{-1}}$. At this altitude, the thickness of the glacier is around 600 m, its surface slope is approximately 0.032, and its depth-averaged velocity is approximately $1.3\ \mathrm{km\ a^{-1}}$. Hence, in a year, a unit cross-sectional area column of ice would descend around 40 meters, releasing approximately 2.2×108 J of potential energy, enough to melt approximately 0.7 meters of ice (Björnsson, 2003). While some of this melting will occur inside the body, the majority will occur close to or at the bed. Therefore, at this elevation, bottom melting could account for up to 14% of the overall ice loss. However, on the majority of glaciers, this type of melting represents a significantly lower portion of the overall volume and may be disregarded in mass balance calculations.

6.6 ANTARCTICA

More than 90% of the ice loss in Antarctica occurs from calving, with ice shelves accounting for the majority of this loss. Compared to tidewater glaciers, these calvings typically discharge much larger slabs of ice. This is most likely a result of stronger ice shelves. They would be stronger due to their lower temperatures and less severe crevassing. An iceberg width, measured normal to the calving face, is probably comparable to the shelf thickness. But many icebergs are considerably larger. In April 2000, Iceberg B-15 separated from the Ross Ice Shelf in Antarctica, with dimensions of 37 × 290 km and a thickness of around 430 m (Lazzara et al., 2008). The mechanisms generating ice bergs of this magnitude remain inadequately understood. Recent research has revealed that polar ice shelves can disintegrate quickly. In 1995, the 1600 km^2 Larsen A ice shelf broke up in 39 days, while in February 2002, the 3250 km^2 Larsen B shelf broke up in just 41 days (Cape et al., 2014). There appears to have been substantial melting on the shelf surface as a result of climate change. It appears that the crevasses spread over the shelf, causing the breakdown. Periodically, calving during the Wisconsin glacier created icebergs that discharged coarse sand as they floated across the North Atlantic Ocean. It is generally accepted that the partial collapse of the Laurentide Ice Sheet over Hudson Bay and the swift outflow of ice across the Hudson Strait were caused by these ice-age calving episodes. It is up for debate as to whether they were caused by the collapse of an ice shelf that supported the Labrador Sea or if they were solely the result of tidewater glacier retreat. Under floating ice shelves, ocean currents pierce through, mixing salty seawater with freshwater that subglacial drains from the ice sheet interior. Depending on the temperature and salinity of the combination, the pressure, and the temperature differential in the basal ice, melting or freezing may occur at the foot of the ice shelf. It is true that melting can occur in one location while freezing in another. Up to 20% of the mass loss in Antarctica may be attributed to bottom melting beneath ice shelves.

6.7 CLIMATE SHIFT

The consistency of glacier mass balance patterns varies on spatial scales. Global climate shifts are one thing; these include the enormous ice advances of the Pleistocene and the smaller ice advances of the Little Ice Age. These are both well-known and little understood, with the exception that the longer cycles appear to have been controlled by changes in the Earth's orbit that impacted the timing and amount of solar radiation received at higher latitudes. Small-scale regional weather changes could lead to distinct behaviors of glaciers that are just a few hundred kilometers apart. The net balances of Alaska oceanic glaciers typically do not match with those of glaciers in southern Canada and neighboring areas of the United States between the mid-1960s and the late 1980s. Storms are diverted toward Alaska, resulting in high winter balances, when the Aleutian Low, the atmospheric low-pressure system off the Aleutian Islands, is normal in the fall and winter. Nevertheless, storms continue further south and accumulation rates are high in British Columbia and Washington when this low is not as deep as it usually is. The mid-1980s saw the breakdown of this pattern. Since then, the South Cascade and Wolverine glacier winter balances have remained out of phase, due to a sharp rise in ablation, both glacier net balances are now negative, indicating that net balances are in phase. The summer low along the west coast also affects summer balances in western North America. Large negative summer balances result when conditions are hot and dry and there is a similar high over British Columbia, which is when this low is relatively deep.

6.8 STORMS

The size of pressure patterns can also lead to asynchrony in mass balances. Cyclonic storms with counterclockwise winds occur in the winter as a result of small-scale low-pressure disturbances, which are distinguished by changes in the height of the 500-mbar surface (the surface on which the atmospheric pressure is 500 mbar, or roughly half the pressure at Earth's surface). These storms raise the winter balance on Sentinel Glacier (British Columbia) and are linked to large-scale air patterns. On the other hand, regular high-pressure disturbances that produce anticyclonic patterns and, consequently, clockwise winds, reduce accumulation in the winter and enhance melting in the summer. In contrast, only more important disturbances associated with long-wave patterns over the North Pacific have an impact on Peyto Glacier, which is located roughly 500 km east of Sentinel Glacier. The El Nino–Southern Oscillation, or ENSO, is one of the most significant of these. In the equatorial Pacific, winds blow westward under normal conditions.

This causes the ocean surface current to flow westward, raising the sea surface elevation in the western Pacific in comparison to the Peruvian coast. The westerly wind current weakens, the upwelling is reduced, and the ocean and consequently the currents off Peru warm at intervals of two to six or seven years. Weakening the westerly wind flow even further, the warmer air lowers the pressure gradient between Peru and the western Pacific. As a result, an area of intense precipitation that is found in the western Pacific moves eastward. El Nino conditions eventually subside, and normal or even cooler conditions emerge. Although the exact cause of El Nino remains unknown, its effects are extensive, impacting not only the glaciers on the northwest coast but also weather patterns in the Pacific, North America, and far beyond.

6.9 OCEAN-ATMOSPHERE INTERACTIONS

Despite its global influence, ENSO is caused by ocean-atmosphere interactions that occur internally in the tropical Pacific and the surrounding atmosphere (Nur'utami & Hidayat, 2016). We now know that the ocean and atmosphere also experience similar oscillations. The first is the PDO or Pacific Decadal Oscillation. Sea surface temperatures or SST in the eastern equatorial Pacific are somewhat warmer than average during the warm phase of the PDO, but they are noticeably cooler in the northwest Pacific. There appear to be two predominant periodicities in the PDO: 15–25 years and 50–70 years. There are sudden shifts from one warm phase to the next. The warm phase started in 1977 and seems to have ended in 1998. El Nino often intensifies during the warm phase of the PDO and becomes normal during the cold phase. During the positive phase of this oscillation, the jet stream is farther north and a strong low-pressure area is present across southern Greenland and Iceland in the winter. As a result, North Africa is drier and North Europe is warmer than usual. This might have an impact on the Alps and Scandinavian glacial mass balance. Periodicity of 2, 8, 24, and 70 years has been found in the North Atlantic oscillation. Temperature and accumulation trends in Antarctica seem to reflect upper atmospheric processes on a larger scale. Heat cannot reach the higher stratosphere due to the warming of the troposphere, the layer of air between the Earth's surface and around 11,000 meters. The stratosphere cools and gets denser as a result, which intensifies downwelling radiation over the South Pole. As a result, enhanced downwelling radiation from the stratosphere cools the interior of the continent as the troposphere around it warms, and vice versa. Clearly, regional differences in mass balance occur on a range of various geographical and temporal scales as a result of atmospheric circulation patterns that we are only now starting to understand. The data base required to recognize and analyze these circulation patterns is growing quickly, and as glaciologists and meteorologists start to build and utilize it.

6.10 GLOBAL WARMING-WHAT IS THE ROLE?

The subject of whether global warming is causing enough ice to melt to account for the observed rise in sea level worldwide is currently of great interest. Currently, the range of 1 to 2 mm a^{-1} is the best estimate for this increase. Uncertainties are high because estimating the mass change of a glacier or ice sheet requires computing the small difference between the total accumulation and total loss. In fact, we are still unable to determine if the Antarctic and Greenland ice sheets were expanding or contracting, despite the estimates currently available. Smaller glaciers and ice caps, on the other hand, have less uncertainty, which implies that 15% to 25% of the increase in sea level may be caused by the melting of these ice masses. There was a mass loss from Alaskan glaciers alone in the late 1990s estimated to be −96 ± 35 km3 a^{-1}, which is higher than the estimate for all valley glaciers and ice. The ocean thermal expansion, permafrost melting, sedimentation, and terrestrial storage in lakes and groundwater reservoirs are other factors that contribute to the increasing sea level. Despite the fuzzy mass balance data for Greenland, which results in a net balance of -44±53 Gt a^{-1}, the pattern is interesting (van Angelen et al., 2014). The observational data indicates that as warmer air brings more moisture inland, the ice sheet is getting thicker in the interior, at least locally. It is, however, weakened (thinned) near the borders where melting is accelerated by rising temperatures. Out of the nine crystalline forms of ice that are known to exist, seven of them are only stable at pressures higher than 200 MPa. We do not need to be concerned about these eight forms because the maximum pressures and lowest temperatures found in glaciers on Earth are approximately 40 MPa and −60°C, respectively.

It is difficult to determine the rate-limiting mechanism. This is since it varies between materials and that it fluctuates within a material in response to stress, temperature, and the presence of impurities. The rate-limiting mechanism in polycrystalline aggregates and single crystals may differ. In unstressed crystals, the dislocation density is typically low and gradually grows to its steady-state value, with the application of stress. This provides more evidence in favor of the theory that drag is the process that limits the rate. However, one more dislocation process needs to be taken into account. Gridlock usually occurs when dislocations start to interfere with one another. These knots of dislocation prevent deformation. Under these conditions, an unrestrained sample that is subjected to stress has a gradual decrease in deformation rate, a phenomenon known as work hardening. Diffusion of atoms away from dislocations is one such process that causes the dislocation to migrate from one crystallographic plane to another. If the dislocations are of the edge type, this is known as dislocation climb; if they are of the screw type, it is known as cross slip. After some initial time, both mechanisms release knots and permit deformation to proceed at a roughly constant rate.

6.11 GRAIN BOUNDARY

Grain boundary movement is thought to be a recovery process that occurs, especially under low stressors, when some crystals develop at the expense of others. Ice behind the boundary is unstressed and hence is free of dislocations. Other mechanisms might, however, also be at play because the difference between these two activation energies appears to be more than that which can be accounted for by experimental error. To maintain the continuity of the medium, neighboring crystals in a polycrystalline aggregate must also deform when one of the crystals in the aggregate does. Thus, the deformation must involve all of the crystals. The majority of crystals are oriented so that a component of the applied force is parallel to the basal plane. Some crystals, on the other hand, won't be oriented in this way, and as a result, they must deform in a direction that isn't parallel to the basal plane. It is possible to slip on the pyramidal planes. On the prismatic planes, on the other hand, the slip would most likely be parallel to the basal plane, which would prevent it from accommodating stresses that are typical of that plane. Because such slip is probably going to be considerably harder than that on other planes, it may be rate-limiting. As a result, slip on pyramidal planes may be required. An ice sample first experiences elastic deformation when it is loaded. At that point, positively oriented grains start to slip on the basal plane and distort. As a result, grains that are not as favorably oriented receive the load, and large stresses may arise in those grains. Peak stresses in individual grains may be many times greater than the sample mean stress as deformation proceeds, indicating an increasing degree of inhomogeneity of the stress distribution.

Experimental research has revealed that monocrystal resistance to shear on basal planes is significantly lower than that of other crystallographic planes. This is due to the previously indicated lack of associations between basal planes. For instance, a monocrystal oriented for easy glide needs a stress of roughly 0.03 MPa to create a strain rate of 1.0 a^{-1} at $-10^{o}C$, while a monocrystal orientated for hard glide needs a stress of almost 3 MPa. Thus, the local stress can vary inside a polycrystalline ice sample. The strain rate of the polycrystalline sample falls between the monocrystal easy-glide and hard-glide orientation values (Castelnau et al., 1998). Grain orientations that are not optimal for basal glide may accommodate some of the plastic deformation in adjoining crystals by deforming elastically as a result of this inhomogeneous internal stress distribution. Thus, elastic strain energy has been stored in these crystals. This elastic energy causes back stress to be applied to the crystals that were previously plastically deformed upon release of the load, leading to a minute amount of reverse creep. As opposed to what would occur in a purely elastic medium, reverse creep requires the grains that were initially oriented favorably for basal glide to creep backward to release the elastic energy that has been accumulated. Ice crystals found in glaciers differ in size and degree of interlocking. For example, if there was no bonding across grain boundaries, some polycrystalline ice samples would hold together, while others would break up into a pile of grains, up to a few millimeters. Texture is the term we will use to describe these features of crystal size and shape. The rheology of ice is affected by both fabric and texture. Like petrologists, glaciologists employ thin sections to examine these processes. The thin pieces are usually 60–80 mm wide and somewhat less than a millimeter thick. The crystal structure rotates the light as it passes through the crystal, and the amount of rotation depends on the orientation of the crystal. This allows for the observation of individual crystals when polarized light is passed through a thin section and then observed through another polarizing filter oriented at right angles to the first.

6.12 GLACIAL MOVEMENT

Early on in the history of glaciology, fixed points on stable surfaces off the glacier were often used as triangulation sites for measuring velocity. Electronic theodolites with lasers significantly decreased the amount of work required to measure in the 1970s. Theodolite activation and stake tracking are now possible with computer-aided devices because of later technical advancements. Measuring can therefore be done automatically as often as needed. With theodolite-based methods, measuring velocity distant from stable fixed sites is nearly difficult. That issue was resolved with the introduction of geographical positioning systems (GPS) technology. GPS devices calculate the distance to a collection of satellites whose orbits are quite precisely known. A point location can be found within a few centimeters using the most accurate GPS devices. Currently, rates of continental drift are measured using these units. Using satellite images recorded at intervals of days to a year or so to establish the velocity is an interesting breakthrough that is rapidly contributing to knowledge of flow in large ice sheets. Based on the presumption that some features are immobile or moving more slowly than others, the images are coregistered. After that, crevasse patterns are compared using cross-correlation techniques to ascertain how much they have moved between the dates of the corresponding images. These comparisons are performed by digital computers operating at high speeds. A thorough quantitative map of the flow can be created in this way. The measurement of velocities, which was previously limited to theodolite precision surveying, is being completely transformed by a method known as satellite interferometry. Flowlines typically point upward in the ablation area and downward in the accumulation area. They are parallel to the bed at the bed and parallel to the surface at the equilibrium line. The flowline descends straight until it reaches the bed. From there, it follows the bed, staying precisely parallel to it if melting or refreezing occurs, emerging at the edge. Throughout the entire length of these flowlines, the volume of ice moving between them stays constant in the steady state. Because it is assumed that the ice is incompressible, the result is that the velocity will be highest where the flowlines are closest to one another. Therefore, the highest velocities will be close to the equilibrium line. There may be noticeable seasonal and geographical changes in the velocity field, especially on valley glaciers. Many studies have demonstrated that glaciers accelerate in the summer when surface water can reach the glacier bed. This occurs as a result of rising water pressure, which raises sliding speed. Even the surface speed varies throughout the day. Seasonal accelerations are higher here because sliding speeds are highest below the glacier midline. Both at the surface and at depth, these accelerations cause detectable changes in the magnitude and direction of velocity vectors. On the other hand, researchers found no seasonal fluctuations in their study using boreholes on the Worthington Glacier, Alaska (Harper et al., 1998). Nonetheless, they discovered prominent spatial fluctuations. There were fluctuations in horizontal speed in a transverse cross-section of the glacier of up to 5%. These changes did not appear to be due to side drag, but they might have been brought on by pressures originating in an ice fall higher on the glacier. The ablation rate is often roughly constant across the glacier. It might be a little higher close to the edges due to heat advected or radiated from the valley. Across the glacier, the longitudinal surface slope will likewise remain roughly constant. The transverse surface slope would produce a transverse component of flow toward the center of the glacier. Pilots flying over Greenland and Antarctica before World War II discovered that their radar altimeters were providing inaccurate information. After further examination, it was found that the radar waves were reflecting off the bed and going through the ice sheet. Thus, the device for radio echo-sounding glaciers was created. The main goal at first was to know how thick the ice was since glaciologists could only previously do this with gravity measurements, seismic profiling, and drilling. Nevertheless, it was quickly ascertained that imaging the inside layers of glaciers and ice sheets was also being done. Exceptionally clear pictures of this layering are now regularly produced by varying the radio wave frequency and using return signals. Subtle differences in the dielectric constant are the source of reflections. Because they move over uneven substrates, glaciers have uneven surface profiles. Furthermore, uneven nunataks may have an impact on their transverse flow patterns. As a result, during the melt season, drifting can cause uneven patterns of accumulation and ablation. Dirty ice with extremely varying quantities of detritus lies beneath the till.

6.13 GLACIAL DEBRIS

Another geomorphic feature that results from the breakdown of debris-covered ice masses is the ice-walled lake plain, which is mostly found in the upper midwest of the United States and nearby regions of Canada. When the ice surface developed depressions due to unequal melting, the lakes would have initially formed. Surface water that warms to +4 °C sinks because this is the temperature at which water is densest(Maslin et al., 1995). Convection that followed would have transported +4 °C water to the lake bottoms, where it might have melted ice if the ice was not too cold and the sediment layer in the lake bottom wasn't too thick. Thus, the lakes would have become deeper. According to field data, several of them eventually made it inside the bed. Subsequently, sediment from supraglacial streams entered the lakes, creating typical lacustrine deposits that included lacustrine silts and clays farther out from the coast and deltas of sand and gravel close to stream mouths. These deposits remained as flat-topped hills in the landscape as the surrounding ice melted. Glaciologists realized in the middle of the 1980s that the flowlines in big ice sheets were not as uniform as they had previously thought.

CHAPTER 7

GLACIER PHYSICS

Contents

7.1 INTRODUCTION

The process of creating glacier ice, which is frequently an intricate one, begins with snow falling on a glacier. Temperature affects both the process and duration of snow-to-ice conversion. In temperate locations, where melting and refreezing of wet snow alternate, snow turns into ice considerably more quickly on glaciers than in central Antarctica, where constant low temperatures prevent freezing. As a result, we need to understand various processes rather than a single transformation mechanism. Depending on how much melting occurs, we must categorize glaciers and even distinct sections of the same glacier into distinct groups. Material that hasn't changed much since it fell is referred to as snow. Materials that are undergoing intermediate stages of transformation will be referred to as firn. Firn originally meant wet snow that remained through one summer without turning into ice. This restrictive meaning is no longer valid since firn can also refer to modified snow on polar glaciers that do not melt. However, the broad definition suffers from the uncertainty that we accept on the distinction between snow and firn. Sometimes, we might use the word snow when firn would be a better choice. Since ice is typically close to melting temperature, unlike other substances, changes in crystal size and structure occur easily. As a result, molecules can travel about rather freely on the crystal surface (surface diffusion) as well as inside the ice lattice (volume diffusion).

7.2 PARTICLE AND ENERGY INTERACTIONS

Furthermore, sublimation occurs easily. It can refer only to the transition from the solid to the vapor phase, or it can be used to describe the entire process of transforming the solid to the vapor, moving through the vapor, and then returning to the solid state. The thermodynamics law states that a system's free energy tends to a minimum, and is reflected in the net direction of movement of molecules. The flattening of the surface and reduction in surface area both lower the free energy. As a result, the molecules tend to redistribute themselves, decreasing the crystals' overall surface area and flattening them out. New snowflakes lose their intricate forms and become spherical particles (Seifert et al., 2019). This is further aided by the snowflakes breaking as they hit the surface or if they are blown along afterward. Since this further decreases the surface area for a given mass, the larger crystals typically develop at the expense of the smaller ones. The movement of individual particles in relation to their neighboring or adjoining particles is a significant element in the early phases of transformation. This is made easier by the particles being rounded. Along their boundaries, the particles glide past each other. The porosity, or the ratio of space between spheres to total volume, is equal to 26% in the rhombohedral structure, which is the closest feasible packing of spheres. However, some studies demonstrate that porosity can never be lowered below approximately 40%. A 40% porosity translates to a density of 550 kg m^{-3} for ice spheres with a density of 910 kgm^{-3}.

The process of transferring material to the places of contact between particles to form bonds might further lower the total surface area. According to laboratory tests, sublimation predominates in the early phases rather than diffusion. This is expected given the high vapor pressure of ice. The uppermost layers of firn, with densities of 350 to 400 kgm^{-3}, may normally be sliced into coherent blocks. Bond formation usually starts before the density reaches 550 kgm^{-3}. Sublimation decreases significantly as density rises and firn becomes less porous. The area of contact between grains and the load both increase at the same time. The two main processes are recrystallization and deformation. Molecular diffusion modifies the size and shape of crystals to lessen the pressures on them and movement of individual crystals along internal glide lines causes deformation. Crystals can easily bend in response to load pressure when there are glacier flow-related deformations present, such as stretching or compression along the glacier. Accordingly, firn densification advances more quickly when the glacier compresses or stretches. At around 730 kg m^{-3} of density, the densification rate slows once again, albeit in certain profiles it is difficult to see this shift. All of the air that is still present at this time is concentrated in narrow channels at the grain boundary intersections. Beyond this stage, the majority of the densification is explained by ice deformation or creep.

7.3 GRAIN TRANFORMATIONS

The air spaces between grains close off when the density reaches around 830 kgm^{-3}. The air that remains, around 10% of the volume, is now only apparent as bubbles because most of it has gone to the surface. Glacier ice has replaced the firn. Compression of the air bubbles by the surrounding ice leads to an additional increase in density (Martinerie et al., 1992). Air bubbles allow researchers to study processes like the accumulation of trace gases in the atmosphere as a result of human activity and the variation of atmospheric carbon dioxide concentration with temperature during the ice ages because they preserve samples of the atmosphere at the time of their formation. The air in the bubbles, however, is younger than the surrounding ice because the air in the firn mixes with the atmosphere. Because not all of the bubbles form at the same depth, this generalization is incorrect. Bubble ages range from a few decades to a few centuries at a particular depth below the firn-ice transition. The majority of these procedures are temperature-sensitive. The pace of change differs from location to location. The process of transformation involves constant changes in the tensions between adjacent crystals. The vertical compressive stress is greater than the horizontal components at low densities. However, the total stress distribution approximates hydrostatic as the firn density becomes closer to that of ice. The orientations of crystals are frequently distributed consistently, as confirmed by studies on cores from dry-snow zones. While most grains are between 2 and 5 mm in size, certain crystals can be significantly larger. The density of a layer of hoar ranges from 100 to 300 kgm^{-3}, and it is porous. While some hoar layers form on the surface, others grow beneath the snowpack.

7.4 DEPOSITS

In unconsolidated snow, a combination of sublimation and vapor deposition processes results in the development of a hoar. It is necessary to have a significant vertical temperature gradient because it creates a vapor pressure gradient that propels molecules from warmer layers into colder layers and out into the cold air above the surface. When vapor passes through the snow and condenses on ice-grain surfaces, faceted crystals are formed. Sublimation simultaneously eliminates the ice from the granules' opposing sides. More ice must be lost through sublimation than must be obtained through crystal development to create low-density layers within the snow. Depth hoar forms more quickly in the snowpack due to the convection of air. The temperature gradient's contribution to the system's increased internal energy is indicated by the growth of depth hoar crystals. Large temperature gradients required for hoar formation are produced by several factors. Autumn weather arrives quickly, chilling the top layers while keeping the underlying layers rather warm. The daily cycle of temperature variations near the surface results in large gradient fluctuations that also change direction. The heat from the sun causes the snow immediately below the surface to warm, which causes the vapor to rise out of the snowpack and, to a lesser extent, into deeper layers. Hoar layers are formed in central Greenland during the summer months by this mechanism. Polar glacier layers, which are often only a few centimeters thick, are clearly visible in walls and serve as indicators for calculating the amount of snowfall that occurs each year. Conversely, at certain low-accumulation locations on the ice sheets, firn accumulates into thick layers that last for several years. In the winter and spring, depth hoar forms along the ground-snow interface in snowfields. Depth hoar forms on nearly the whole snowpack in central Alaska. This region experiences higher temperatures and temperature gradients than the polar areas, and throughout the winter, favorable circumstances for the creation of depth hoar persist. Because they serve as sliding surfaces, depth hoar layers are especially dangerous in mountainous regions.

Grain packing remains the most crucial element in the early phases. Grain rounding accelerates with melting because melting occurs first at the edges of the grain. Because the smaller grains often melt before the larger ones, the average grain size increases. Ostwald ripening is the process by which molecules migrate from smaller to larger grains due to variations in surface curvature, even in the absence of net melting (Madras & McCoy, 2004). Furthermore, after melting, grains can refreeze to form clusters. Since grains experience daily cycles of freezing and thawing, they connect most quickly in the surface layers. Meltwater lubricates the grains, accelerating packing and enabling close packing since a water film surface tension pulls the grains together. Therefore, compared to a dry-snow zone, packing might reach a higher maximum density in a meltwater area. Meltwater that has been refrozen additionally accelerates the last phases of transformation. In this approach, air spaces are filled. A significant amount of meltwater can be quickly converted back into ice in the form of lenses and layers of ice by freezing again. Depending on the amount of meltwater in a given area, the time required to accomplish the transition varies greatly. The extreme situation, where snow turns into ice in a single summer, is shown by a superimposed ice zone. The significant variation in densities with depth in the firn on temperate glaciers is explained by the formation of ice layers. Melting and refreezing generate a slight decrease in surface height because they raise the average density of the firn. In the Greenland Ice Sheet percolation zone, this is a crucial mechanism. Even in the absence of changes in the mass and rate of ice accumulation, the surface elevation varies annually due to fluctuations in the amount of melting that occurs as a consequence of summer temperatures. When a force is applied, ice behaves like a thick, viscous fluid; this deformation, known as creep, happens gradually and constantly. In addition to contributing to a glacier's general motion, creep also explains phenomena like the closing of ice tunnels and the movement of basal layers. Fractures can occasionally occur when a significant amount of stress is applied quickly or under continuous tensile stress. Seismic soundings of ice thickness are based on high-frequency sound waves that are transmitted by elastic deformation. Stress and deformation are proportionate in elastic deformation. Perfect plasticity occurs when stress increases gradually without causing any deformation up until it reaches a crucial value known as the yield stress, at which point deformation starts. Although the deformation can continue at any rate, the stress can never be greater than the yield stress.

Polycrystalline glacier ice is made up of numerous separate crystals or grains (Placidi et al., 2006). The creep behavior of the ice is determined by single-crystal characteristics combined with grain-to-grain interactions. The creep is influenced by grain-scale structural characteristics, such as fabric and texture, frequently in ways that are yet poorly understood. Spatial patterns emerge and change in grain scale structures. Since these structures are measurable, they offer important insights into the long-term dynamics of glaciers and ice sheets. Deformation is closely correlated with grain-scale structural processes, such as crystal development and alignment along preferred orientations. Therefore, even though many of the connections between these processes have not yet been fully established, it is beneficial to consider them all. The characteristics of the H_2O molecule give rise to the structure of ice. An electrical polarity is produced, with an excess of positive charge at the proton corners. This polarity causes adjacent water molecules to join together through hydrogen bonding. A material can crystallize either hexagonally or cubically if each atom has four neighbors organized in the shape of a tetrahedron. X-ray diffraction experiments have clearly demonstrated that the structure of terrestrial glacier ice is one kind, with oxygen atoms organized in layers of hexagonal rings. Two closely spaced planes, one corresponding to the interior oxygens and the other to the basal faces of the tetrahedra, are where the atoms are located. Alternate oxygens are found in the upper and lower planes of the ring. The ice crystal structure is made up of stacked ring layers, where each layer is a mirror image of the one before it. This structure is similar to that of metals with hexagonal shapes, including cadmium, magnesium, or zinc. Glacier ice usually has 10^6–10^9 distinct grains per cubic meter (Gomez et al., 1988). Every grain of fully formed ice is one solid crystal. The characteristics of single crystals along with interactions between the grains, combine to form the bulk properties of glacier ice. The processes that contribute to snowfall on a glacier evaporation followed by atmospheric transport, condensation, and precipitation act as a distillation system. The trace amounts of contaminants do have an impact on its physical characteristics and serve as a record of the atmospheric conditions. Moreover, particles may cluster on an ablation glacier surface, which can have a major impact on its ability to absorb solar radiation and, consequently, the pace of melt. The origins of soluble impurities in ice are soil material assimilation into the base layers of a glacier, dissolving of surfaces, and atmospheric deposition. Percolating water eliminates most of the impurity content in temperate ice, which is at the melting point.

7.5 BUBBLES IN GLACIER

There are many bubbles in glacier ice unless the deep ice sheets are in an area with tremendous pressure. Although bubbles are typically spherical in shape, they can occasionally take on elongated shapes due to the surrounding ice deformation. If the temperature reaches the melting point, bubbles may contain liquid water; otherwise, they contain atmospheric gases. The bubble composition in cold polar ice for noble gases and unreactive gases like diatomic nitrogen and oxygen mostly corresponds to the air quantities at the time the ice formed. Gas compositions are changed upon contact with liquid water due to the varying solubilities of the various molecules. Methane and other chemicals may be produced via biogenic activity. When firn converts to ice in polar conditions, the bubbles take up roughly 10% of the volume. The initial gas level is usually reduced by dissolving in the water when the throughflow of melt begins. As the pressure of the ice layer above rises, bubbles in the polar ice sheets contract with depth. At the same time, the gas pressure inside the bubbles rises. The gas reaches a dissociation pressure, which is dependent on temperature and composition. In polycrystalline ice, the distribution of grain sizes and shapes makes up the texture. An effective diameter (D), radius, and cross-sectional area are used to measure grain size. The apparent average grain size in an ice cross-section is smaller than the sizes of whole grains; the cross-section plane does not pass through the centers of the majority of the grains. Grain can occasionally take on intricate and interlocking shapes. Atoms reattach to the crystal lattice in a position of reduced free energy as they recrystallize. The grain boundary passes through the ice. Grain boundary migration is primarily caused by variations in free energy between neighboring lattices, which might be from variations in lattice strain energy or from the curvature of the boundary. Grain growth causes the mean grain size of polycrystalline ice to grow over time in the absence of processes to create new grains. The energy of grain boundaries makes this process thermodynamically preferred.

Grains proliferate due to boundary migration. Normal grain growth, a process that is driven by the curvature of the boundaries in slowly deforming ice, is what drives migration. A net flux of atoms crosses a grain boundary. Smaller grains typically have fewer boundaries and a smaller radius of curvature than large grains. Therefore, the boundary will frequently migrate into the protruding side of the smaller grain, which occupies that space. Through this process, the polycrystal mean grain diameter rises as huge grains gradually consume/use little grains. While temperature has a significant influence on grain growth, contaminants at a grain boundary can lessen or even completely eliminate grain migration. Metallurgical principles were applied to tackle this issue with ice. Researchers concluded that, while they function differently, soluble contaminants, microparticles, and bubbles can all lower grain growth rates. The intrinsic mobility of the grain boundary is decreased by soluble contaminants. Conversely, microparticles and bubbles lessen the effective pushing force. When a grain deforms, it creates additional dislocations and other lattice defects in the crystal that distort the lattice and so increase the energy stored. The lattice strain energy is this additional internal energy component. A net transfer of molecules from the higher-energy grain to the lower-energy one is caused by a difference in the lattice strain energy density between two adjacent grains. As a result, the grain boundary moves into the grain with more energy. When grain boundaries shift freely, as they do in warm ice, migration recrystallization is preferred over polygonization (Faria, 2006). Polygonization, the more likely process at low temperatures, is similarly driven by accumulated strain. When polygonization is sufficiently active, it slows down the rate of grain growth and stops the mean grain size from rising any higher. This could be one reason for the shift to a more consistent grain size. Additionally, nucleation plays a role and gets more active with increasing depth. These changes most likely result from high quantities of soluble and particulate contaminants which restricts grain formation. Because high temperatures accelerate grain development rates and produce slow deformation in some places, such as ice splits and bedrock hollows, coarse grains can form at even deeper depths near the bed. In other places, the ice close to the bed rapidly distorts. Grain sizes vary widely in these conditions, with fine grains typically occurring in layers with calving and high impurity levels. These fabrics are not altered by polygonization, but they are dramatically changed by nucleation and migration recrystallization. Older grains that have orientations that are not conducive to basal glide are removed first. When strain builds up, new grains rotate after nucleating at orientations that are conducive for basal glide. Recrystallization fabrics can form more quickly at high temperature and/or stress conditions, however, deformation fabrics require a finite strain of at least 10%. Single-fabrics typically form in the middle to upper layers of the polar ice sheets, a zone dominated by vertically oriented uniaxial compression. Ice shelves, which experience rapid vertical compression, longitudinal extension, and active recrystallization, are likewise characterized by well-developed fabrics. Unconfined uniaxial compression has also been used in the laboratory to create these fabrics. Fabrics have developed via uniaxial extension in a single horizontal direction in the upper strata of the ice sheet at Vostok, in central Antarctica. Multi-maximum fabrics have been found in polar glaciers, base layers of ice sheets, and temperate glaciers. Although the cause of multi-maximum fabrics cannot be properly described, complex stress perpendicular to the plane and active migration recrystallization and nucleation are likely the causes. A girdle fabric or one with multiple maxima can be produced by such a stress factor. In the lab, multi-maximum fabrics have also been created by repeatedly compressing the same sample and annealing it in between.

7.6 DEFORMATION STUDY RESULTS

Further research is necessary to fully comprehend the potential direct and indirect impacts of impurities on ice deformation. The range of pressures, temperatures, impurity concentrations, and grain sizes that are relevant to glaciers is expected to have a significant impact on the effects. Only the basal planes of a single ice crystal are easily calved. Hence, the c-axis fabric, which frequently exhibits some degree of preferred orientation in glaciers, is a major determinant of the viscosity of polycrystalline ice. This feature is disregarded in almost all theoretical assessments of glacier flow. The effects of fabric are directional, as opposed to temperature-related fluctuations in viscosity; the direction of the deformation determines whether the fabric will soften or stiffen. The fundamental characteristics of anisotropic deformation of ice are known, however no universal creep equation/relation has been developed. When polycrystalline ice is deformed using a tight single-maximum fabric, it should mimic the deformation of an individual ice crystal. The planes that are perpendicular to the fabric axis, or the mean direction of the c-axes, should experience shear. Shear planes tilting 45 degrees from the fabric axis should cause the deformation to become more severe as the shear plane is moved away from this location, eventually reaching its firmest state. Samples with a strong single-maximum fabric were subjected to uniaxial compression, with the compression direction being changed. Measured creep rates varied by more than a factor of 100 from soft to stiff orientations (Castelnau et al., 2008). Compression applied in a plane results in a deformation rate comparable to that of isotropic ice. In contrast, deformation occurs at a rate that is only 0.06 times that of isotropic deformation when compression or extension is applied perpendicular to the layer.

If positioned vertically, a single-maximum fabric amplifies the distortion. This accounts for a significant amount of the shearing of Arctic ice caps. Additionally, Meserve Glacier, an Antarctic Mountain glacier with a cold base, has rapid basal ice shearing due to the presence of extremely strong fabrics. Borehole tilt is increased when shear on horizontal planes is enhanced by a robust, vertical single-maximum fabric. On the other hand, shears at various orientations are used to close a borehole, and the fabric stiffens the ice in this process. As a result, even if the layers with this fabric tilt faster, a borehole closes more slowly in them. The specific recrystallization processes and deformation mechanisms that are at work may determine how anisotropically ice deforms with a particular fabric. However, the assumption that deformation reflects the collective average of features of many individual crystals might explain major aspects of anisotropic deformation. Individual crystals in polycrystalline ice predominantly deform via basal glide. The bulk deformation is therefore governed by the crystals shearing on basal planes, which results in redistributions of stress from crystal to crystal and geometric effects i.e., the orientation of basal planes in response to applied stress. A crystal that is orientated favorably for basal glide is partially protected from stress.

7.7 MASS BALANCE

Studies on mass balance are crucial for a variety of reasons. Glacier mass balance measurements show the net influence on sea level and whether or not there are global changes in the amount of glacial cover on a landscape (Forsberg et al.,

2017). In recent studies of polar ice sheets, measuring the mass balance of a zone has proven to be particularly useful in revealing significant changes due to ice movement. A large portion of the water used for residential use, agriculture, and hydroelectric plants comes from streams supplied by glaciers in certain areas. A glacier functions as a natural reservoir, retaining water in the winter and releasing it in the summer, giving these streams their unique run-off pattern. During hot summers when there is a shortage of water from other sources, very significant amounts are released.

Although the meltwater refrozen in the seasonal snowpack can be quantified, the intricate percolation and drainage processes in the snow cause significant variation at scales of centimeters to meters. This continues to be a significant source of uncertainty in models and mass balance observations for high-altitude and polar glaciers. The chemical and physical properties of ice formed via freeze are distinct from those of the glacier ice that covers it. When subglacial water at high hydrostatic pressure moves into an area of lower pressure, supercooled water freezes. Such water needs to reheat in order to stay at the same temperature as the ice on top of it since the melting point rises as pressure decreases. The water will become supercooled and colder than the ice if it does not warm up rapidly enough. Subglacial lakes, ice shelves, and glaciers with deep beds are all potential locations for supercooling. As one moves down the glacier, the thickness of the ice diminishes, lowering pressure and raising the melting point. In addition, as the bed elevation rises, less potential energy is converted to frictional heat by the water moving down-glacier along the bed. Supercooling can occur beneath a warm-based glacier due to a combination of decreased ice thickness and raised bed elevation. In certain places, a substantial volume of water freezes to the bottom of the ice shelf, while in other places, the melting process is remarkably quick. Where channels of buoyant water rise along the sloping bottom of a shelf, supercooling may occur. In addition, the base of an ice shelf is distinguished by longitudinal ridges rather than being planar. Basal freeze may occur without the need for supercooling. An ice shelf circulation of cold water can protect the ice from adjacent ocean waters and geothermal heat. Since the bottom layers of shelves are often warmer than the layers above, heat conduction upward into the shelf causes a freeze.

7.8 ICE SHELF

The Amery Ice Shelf in East Antarctica is one prominent ice shelf with basal accretion (Galton-Fenzi et al., 2012). Water moves beneath the shelf in a clockwise manner. On the east side, saline water enters the cavity, interacts with melting snow, and becomes fresh water. After that, it rises along the west side of the shelf base and returns to the open ocean, where it freezes and supercool.

On the western portion of the shelf, the thickness of accreted ice ranges from 50 to 150 meters, with a maximum of roughly 190 meters. Under the Ronne Ice Shelf, one of the biggest shelves in Antarctica, basal accumulation is also notable. Often, accumulated ice melts again before it gets to the shelf. However, in places where it does survive, organic material from the initial saltwater occasionally results in icebergs. Radar observations of ice-thickness changes can be used to predict melt rates beneath the ice shelf. Melt rates are typically, however, obtained indirectly. A glaciological mass balance is one approach that calls for surface balance measurements, ice inflow and outflow data, and shelf shape changes. Using a combination of oceanographic and glaciologic methods, researchers estimated the basal melt rates of Pine Island Glacier, one of the main glacier streams in the Amundsen Sea region. A mechanism is required for subshelf melting to mix and drain the water and specifically to pull deep or intermediate waters into the subshelf cavity. Convection can be fueled by buoyancy. The features of the surrounding water masses and the shape of the basin beneath a shelf are significant variables. Deep water has the warmest water masses surrounding Antarctica. They start when salts are concentrated during the creation of sea ice on the continental shelf, resulting in water that is denser than that of colder surface waters close to the coast. Depending on where they originate, intermediate and deep shelf waters in Antarctica can have temperatures between -2 and +2 °C. Since thick ice shelves are exposed to these deeper and intermediate waters more than thin ones are, they usually melt more quickly. The process of separating ice blocks from a glacier's edge is called calving.

The calving rate is likely influenced by several variables, but no predictive relationship integrating them has been developed. The frontal cliff collapses due to the weakening of the terminus, therefore the rate of ablation below the sea level may be significant. The quick melting of ice at the front is caused by both the mechanical impact of waves and the heat carried by currents. Large icebergs, however, are produced by many swiftly moving and floating glaciers, and their source must be a fracture deep inland of the cliff. Tensile failure is brought on by the longitudinal and transverse strain, which encourages fracturing. The stretching typically gets worse when a glacier moves more quickly. The relationship between the stretching rate and tensile fracture depends on the ice stiffness, which is temperature-dependent; colder ice fractures more slowly because it is stiffer. Thick ice prevents total fracturing, although surface melting water can fill in cracks. Fractures can also be influenced by stresses linked to buoyant forces and cliff morphology that are only significant at the ice front. All of these variables can, in theory, be linked to more variables like the flow regime, tides, sea level, ocean bathymetry, and air and water temperatures. Because of the North Atlantic supply of moisture and ongoing cyclonic storm activity, precipitation happens all year round (Pérez-Alarcón et al., 2023). As temperatures drop and the distance from the main oceanic moisture supply rises, precipitation often decreases from south to north. Melt rates in the south can surpass 5 myr^{-1} close to the edge. The surface meltwater runoff eliminates about half of the annual accumulation. Runoff exceeds sublimation losses by approximately a factor of four. About 50% to 60% of total ablation near sea margins is accounted for by calving and melting.

7.9 CLIMATE

All available surface and upper-air data are incorporated into reanalyzed climatology, which offers the best estimate for atmospheric conditions throughout time. This helps to constrain and direct model-based daily meteorologic reconstructions. The regional model uses this forcing at the boundary to estimate the surface energy balance, snow-rain partitioning, and precipitation everywhere on the ice sheet. The energy balance then produces estimates of sublimation and melting. The outcomes were adjusted based on observations of weather patterns and surface balances from a network of ice sheet stations. Thus, one way to think about the regional weather model is as an interpolation system based on physics. The surface balance is dominated by snowfall and melting. A little over 34% of the melt refreezes, leaving it out of the ice sheet outflow. About 10% of the snowfall is eliminated by direct sublimation. The sublimation of blowing snow is half that of direct sublimation. The Antarctic Peninsula is heavily impacted by the surrounding ocean and reaches lower latitudes than the rest of the continent.

The most efficient time to absorb energy is after the insolation peak since albedos on snow and ice decrease with time throughout the melt season. At the solstice, insolation reaches its peak, but glaciers are still mostly covered in winter snow. Darkening of the surface increases the absorbed or net shortwave radiation available for melt by a factor of three to four as the melt season goes on. For this reason, peak melt rates on mid-latitude glaciers happen one to two months after the solstice. Albedo, for instance, is influenced by the angle at which descending radiation is incident. This indicates that the albedo is dependent on the surface microtopography, the amount of direct and diffuse radiation, and the time of day. Since thin snowpacks are transparent, the underlying material affects the albedo to some extent. Enough sediment buildup on the surface of a glacier protects the underlying ice and slows down melting. The reflectivity of snow and ice changes with wavelength, reflecting visible wavelengths more than near-infrared wavelengths. In the atmosphere, there are additional wavelength-dependent processes. When humid, a larger proportion is composed of visible wavelengths due to the preferential absorption of near-infrared wavelengths by water vapor and droplets. When humid conditions begin, the albedo should rise because the surface reflects more visible wavelengths. An easy substitute for energy-budget calculations is to use empirical regression equations that relate reported ablation rates to air temperatures and other variables recorded by weather stations. The weather stations might be close to or on the glacier. These equations can be used to estimate ablation on glaciers without data, anticipate the rise in ablation rate with warming, and extend a sequence of annual ablation measurements backward in time. These findings must be viewed as imprecise approximations. Limits on each element of the energy budget might be used to evaluate uncertainty.

7.10 ICE MELTING

The main sources of water include runoff from valley sides in mountainous regions, rain, melt from the surface, and melt along the bed and within the ice. By far the greatest source of glacier formation is surface melt, which is completely absent from much of Antarctica and the center of Greenland and often weak in accumulation zones. The amount of water grows and water pressures rise when drainage is unable to completely remove all of the water that enters the bed. The water pressure lessens the forces of contact between the ice and the underlying rock by partially offsetting the weight of the glacier. The majority of drainage happens at the ice-substratum interface. Ice tends to fill gaps by flowing into them. Significant water pressure is not necessary for drainage. Concentrated flows melted ice where abundant amounts of surface water reached the bed, expanding passages and forming an effective drainage system. The basal slip motion of a glacier is influenced by its basal water system and vice versa (Howat et al., 2008). Water on the bedding lubricates the surface and makes sliding easier. High water pressure levels and the ice separation from its substrate are the causes of lubrication. Both make the interface more prone to slipping. Elevated water pressure also weakens basal sediment shear strength, which is necessary for their deformation. On the other hand, bed deformation may block channels carved into sediments, and sliding over an uneven rock bed can also reveal caves that store and transport water. So, the relationships are reciprocal: hydrology affects slide, while slip affects hydrology. Though their exact nature is still unknown, these feedbacks are crucial to the occurrence of events like glacier surges and seasonal fluctuations in glacier flow. The basal water system varies dramatically over time in various locations. Water pressures and fluxes under melting glaciers change significantly throughout the year and in daily cycles. Therefore, a network of extensive passages that allows for a free hydraulic connection between the bed and the glacier surface must exist.

CHAPTER 8

GEOSPATIAL TECHNIQUES FOR GLACIER SCIENCE

Contents

8.1 INTRODUCTION

Remote sensing is the process of gathering data from an object without coming into close physical contact with it. The term is typically used in a narrow sense, where the information is carried by electromagnetic radiation, such as visible light, infrared or ultraviolet radiation, or radio waves, and the observation is made from above the object of interest using a sensor carried on an airborne or spaceborne platform. The term passive refers to a form of remote sensing where the radiation comes from natural sources, or active where it comes from the sensor transmitting the radiation to the object being studied. Originally derived from aerial photography, passive remote sensing can be understood as an expansion of the concept of aerial photography to include additional regions of the electromagnetic spectrum, alternative technologies for radiation detection and data storage, and alternative platforms for the sensor. The invention of radar for military purposes during World War II gave rise to active remote sensing. The utilization of spaceborne platforms has been one of the most important reasons for the increasing application of remote sensing to many studies in the environmental sciences and other fields. The Landsat I satellite launch in 1972 marked the official start of the satellite remote sensing era, even though remote sensing equipment was launched into Earth orbit in the 1960s (Cracknell, 2018). When comparing spaceborne remote sensing to aerial observations, there are several benefits. Even though a single satellite mission's projected lifetime is often no more than three to five years, this is still a far longer duration than what an aerial platform can achieve for continuous data collection. Nonetheless, missions are frequently planned to offer consistent data coverage that continues for decades, if not longer, from earlier missions. Spaceborne remote sensing is especially useful for studying glaciers because of all these benefits, and the glaciological research community was quick to recognize and take benefit of the opportunities presented by satellite data. The terms thermal radiation and black-body radiation are used to describe the natural source of EMR. This is the radiation that all bodies with temperatures higher than absolute zero, or all bodies, emit. The wavelength-dependent distribution of this radiation is dependent on the body temperature. For instance, the predominant wavelength of an object at 0°C (273 K) is approximately 11 μm, and the majority of normal land temperatures produce radiation in this spectrum. For this reason, the region in the infrared spectrum that falls between approximately 3 μm and 15 μm is sometimes called the thermal infrared region.

8.2 REMOTE SENSING

At wavelengths of approximately 20 μm to a few millimeters, the atmosphere is largely opaque; but, at longer wavelengths, in the microwave range, detectable levels of thermal radiation can be found. Passive microwave remote sensing is based on this. The basic principle of remote sensing is that EMR carries information about the object under investigation. This involves assessing a few of the physical attributes of the EMR. The fundamental characteristic of passive solar radiation detection systems is the brightness of the radiation that reaches the sensor. Radiance is a measurement of the radiation intensity in a certain direction that is frequently expressed as a function of wavelength. In a typical remote sensing device, radiance is measured for many distinct wavelengths that span the visible and near-infrared spectra. The reflectance of the surface from which the radiation was reflected is a derived quantity that is frequently calculated using the radiance. This is dependent on the extent of the observation as well as the amount of radiation that was incident in the same wavelength on the surface of the Earth. Atmospheric correction techniques like MODTRAN are required to transform the at-satellite radiance to an at-surface radiance (Chatterjee et al., 2017). Nonetheless, expressing the radiance in this instance as a brightness temperature is also helpful in studies. This is the temperature at which a black body would output the same amount of radiation. Differentiating between at-surface and at-satellite brightness temperatures is important, just like in measurements of reflected solar energy. Emissivity is the difference between the actual (at-surface) radiance and the ideal body emitting radiance at a certain physical temperature. The highest value of emissivity is 1, as no object can radiate more than a perfect emitter at the same temperature. It is possible to calculate the physical temperature of a body whose emissivity is also known by knowing It is at-surface brightness temperature.

Because the characteristics of the light can be modified, active remote sensing systems can offer greater flexibility. Active instruments can be broadly classified: ranging instruments, which measure the distance between the instrument and a reflecting interface, which could be the surface (glacier) or a sub-surface interface (interface between a glacier and the underlying bedrock). Other types of active instruments include those that measure surface reflectance. There is no hard and fast rule in this distinction; instruments can incorporate both elements. There are ranging instruments that use both the radio and visible portions of the spectrum; however, in every case, the main variable that needs to be measured is just the time it takes for an EMR pulse to travel from the instrument to be received back. In general, active surface reflectance measuring devices are all forms of RADARs (RAdio Detection and Ranging) since they function nearly exclusively in the microwave region of the spectrum. The backscattering coefficient, expressed in decibels (dB), is the variable that these devices often measure, even though the idea and surface reflectance are extremely similar. The backscattering coefficient is frequently influenced by radiation. Determining the backscattering coefficient can also be influenced by the polarization of the radiation that is emitted and received. In this case, the radiation that the sensor transmits and receives is referred to as polarization. Both horizontal (H) and vertical (V) polarizations are possible for signal transmission and reception. Four possible combinations of polarization are HH, VV, HV, and VH. In the case of radar interferometry, the phase of the returning signal may also be significant. Radar uses EMR wavelengths that are generally centimeters long, much longer than those found in the visible and infrared regions of the spectrum. These wavelengths are commonly called after different alphabetic characters, such as K, X, C, S, L, and P. Alternatively, radar radiation frequency can be used to identify the region of the electromagnetic spectrum it occupies. This is especially helpful because, as radiation travels through materials with varying refractive indices, wavelengths change but frequencies stay constant. The radar wavelength is a crucial factor in the detection of many ground-based phenomena, much as optical measurements.

8.3 ANALYSIS STUDIES

One of the most vital metrics that may be used to evaluate the strength of an instrument for its intended application is its resolution, or resolving power, which comes in second after wavelength and measurement type. This has to do with the instrument's capacity to discriminate between two comparable objects. The four aspects of resolution are spatial, spectral, radiometric, and temporal. The ability to recognize spatially distributed features in an object or to discriminate between a point object and a horizontally extended object can be understood as spatial resolution in this context. It is commonly believed to roughly correspond with the size of a pixel, or more accurately, the size of the surface element/object that is captured in a single pixel (picture element). Although this is a bit of an oversimplification, unless the imagery has been greatly oversampled (in which case the resolution will be coarser than the pixel size), it is frequently a good approximation. In analog (film) photography, the resolution is typically specified by specifying the photograph scale because the image is not represented by pixels.

8.4 SPECTRAL RESOLUTION

Since scanners are being used more and more to digitize aerial photos, a useful way to describe their spatial resolution is by measuring their pixel size, or ground resolution. A higher level of spatial resolution leads to an increased capacity to identify various surface features. For imaging systems other than synthetic aperture radars, the spatial resolution can be stated as the product of the range i.e., the height above the surface for a downward-looking sensor and the angular resolution. The diameter of the objective lens or, in the case of microwave systems, the antenna and the wavelength of the radiation both essentially define the angular resolution; the finest angular resolutions require large lenses and short wavelengths. As a result, the same instrument will produce a spatial resolution proportional to the height when it is placed at various altitudes above the surface of the Earth. This is very important when deciding between an airborne and a spaceborne system. While satellites used for the same purpose typically orbit at least 700 km above the surface, aircraft used to collect remote sensing data often fly one or a few kilometers over the surface (Brown et al., 2005).

Thus, the greatest spatial resolutions that are typically provided by airborne systems are typically one hundred times finer than those obtained by spaceborne systems. For instance, aerial imaging systems can provide resolutions of tens of centimeters, whereas spaceborne imaging systems operating in the visible/near-infrared region of the spectrum often produce resolutions of tens or hundreds of meters. However, with the introduction of very high-resolution satellite imagery, like that from IKONOS, QuickBird, or EROS, which offers spatial resolutions of the order of one meter or finer, this disparity has lately decreased. Since maximizing spatial resolution may result in other constraints, not all imaging systems are made with that in mind. The term spatial resolution also refers to range resolution when referring to ranging devices such as ground penetrating radars (GPR) and laser scanners, sometimes referred to as LiDARs, or light detection and ranging. This is mostly independent of height and is regulated by the length of the transmitted pulse. The capacity to differentiate between radiation with various wavelengths is referred to as spectral resolution. The number and width of the instrument wavelengths, or the spectral ranges of sensitivity, are typically used to describe spectral resolution. Spectral resolutions are higher for narrow bands. As this might be indicative of the surface type or condition, imaging systems operating in the visible/near-infrared region of the spectrum are typically built to record the fluctuation in brightness, and consequently reflectance, with wavelength. To a certain degree, color photography accomplishes this goal by adjusting to the amount of light that falls on film in the red, green, and blue wavelengths. In analog photography, filters are found within the film; in digital camera imaging, they are found within the camera. These spectral ranges are thus determined by these filters.

Each band in the spectral resolution is around 100 nm wide, which is relatively coarse. To achieve what is commonly referred to as multispectral imaging, the majority of visible/near-infrared imagers offer a variety of wavebands, which are typically 50 to 100 nm wide and defined by filters (Tran, 2003). Certain devices are capable of achieving a far finer spectral resolution; these are typically used to resolve specific features like absorption lines. Hyperspectral imagers can provide hundreds or thousands of narrow continuous wavelengths. Although there hasn't been much use of hyperspectral sensors in studying glaciers, some work has been done on snow surfaces. The capacity to differentiate between two comparable but distinct radiations is known as radiometric resolution. It can be conceptualized as consisting of two parts: the number of levels into which this band is divided and the range of quantities of radiation to which the instrument can respond without the response becoming saturated. When it comes to digital data, the quantity of bits utilized to express the brightness value governs the latter. 8-bit data is frequently used in digital imaging systems because it is easily compatible with a wide range of computer-based processing systems. This means that the range between the minimum and maximum values that can be represented is divided into 255 steps/stages (digital numbers, DN). The value of the radiance, or any other parameter describing the radiometric response, such as the brightness temperature or the backscattering coefficient, is represented by an integer between 0 and 255. But occasionally, the number is less than 255 since some numbers might be set aside to denote special circumstances, like the fact that the data are unreliable for whatever reason. For instance, information from the ETM+ sensor's band 1 on board Landsat 7 is stored as 8-bit integers, which can be utilized to approximate radiometric resolution. Depending on whether the band is set to its low or high gain level, the lowest representable value of the radiance in this band is −6.2 W m^{-2} sr−1μm^{-1}, while the greatest value is either 293.7 or 191.6 in the same units. This means that the radiometric resolution of this band is either 1.18 or 0.78 W m^{-2} $sr^{-1}\mu m^{-1}$, which translates to a reflectance resolution of about 0.2 or 0.1%. For thermal infrared channels, the same rules apply. Using the ETM+ instrument as an example, the band 6 (thermal infrared) data are represented as 8-bit numbers, with the greatest value being 17.04W m^{-2} $sr^{-1}\mu m^{-1}$ and the minimum value being 0 equating to a radiance. For brightness temperatures around 273 K (0°C), this results in a radiometric resolution of 0.067 W m^{-2} $sr^{-1}\mu m^{-1}$, and a comparable resolution in the at-satellite brightness temperature of roughly 0.6 K (°C) (Bhattacharya et al., 2010).

8.5 TEMPORAL RESOLUTION

When data are regularly obtained from the same location, the idea of temporal resolution is typically applied; in this case, it is just the frequency of data collection. Many satellites for remote sensing are positioned in perfectly repeated orbits, where the sub-satellite track orbits the surface of the Earth in a closed curve and returns to the same spot after n1 days and n2 orbits, where n1 and n2 are integers. The ratio of n2/n1 for satellites in low-Earth orbit is almost 14.3. For instance, Landsat 7 has n1 = 16 and n2 = 233 on an identically repeating orbit. Therefore, the temporal resolution of the ETM+ sensor carried by Landsat 7 is 16 days: it is possible to image a specific place on the Earth's surface once every 16 days. The operating agency's decision over whether to gather data during a specific satellite orbit and whether the region is cloud-free will determine whether or not the potential is realized. In contrast, the ETM+ sensor temporal resolution is less than 16 days. The Landsat 7 satellite has a separation of roughly 120 km at a latitude of 45° North or South, between spatially contiguous sub-satellite tracks (Hansen et al., 2016). A specific place on the Earth's surface may be scanned more than once every 16 days due to the ETM+ sweep width. As one approaches the poles, the temporal resolution rises due to the convergence of lines of longitude. The sensor swath width is the primary factor influencing temporal resolution; larger swaths provide more opportunities for observations. It is not possible to choose swath width and spatial resolution entirely apart, and the maximum swath width often equates to no more than 10,000 pixels. This suggests that choosing a spatial and temporal resolution simultaneously is likewise impossible, which is a major justification for the continued operation of lower-resolution sensors alongside better-resolution sensors.

A property of electromagnetic radiation observed at the sensor is what is measured in remote sensing. It is necessary to infer some glacier property from this. In certain instances, a clear correlation exists between the observed characteristic of the radiation and the relevant geophysical parameter. Examples of such scenarios include determining the physical temperature surface from the brightness temperature of the detected radiation. For instance, defining the boundary of a snow-covered region could be done using an aerial photo or a digital image taken in the visible/near-infrared spectrum. This distinction may be made because snow has a higher reflectivity than any other material in the image, allowing the image to be divided into sections with and without snow. Such classification could be done automatically or manually, and it might require training. In that scenario, the algorithm is calibrated using areas that are known to represent snow and non-snow kinds, for example, through field work. Multi-dimensional datasets, such as multispectral data from a sensor like the ETM+, SAR backscattering data from several dates, and so on, are used in more complex classification techniques. There are many commercial and public-domain computer packages available for image processing and classification, including GRASS, ERDAS Imagine, ER Mapper, ENVI, and IDRISI (Brunet et al., 2008). Aerial photography is a well-established method in remote sensing and has been used extensively, particularly in the study of glaciers. Since it makes use of the visible and near-infrared portions of the electromagnetic spectrum, it cannot be done in the dark or through cloud cover. Space images typically have far lower resolution than aerial photos, although having a larger spatial coverage since the spatial resolution of an aerial photograph is dependent on the height from which it was taken.

8.6 PHOTOGRAPHY

Photography is still valuable today, despite the inconvenience of the photochemical processing involved. Digital camera imagery obtained through low-cost systems like EnsoMOSAIC or digital camera systems like Leica ADS40, UltraCamD, or Intergraph DMC is starting to compete with photography in glacier studies as well as other fields. However, because the glacier surface is uniformly bright, one challenge in three-dimensional mapping of glacier surfaces using digital images is that of recognizing surface characteristics. For places where a laser scanner operation would be too complex and costly, especially for smaller glaciers, aerial photography or digital camera imaging still perform better. An aerial picture taken over 6000 meters above sea level on Mt. Kenya in equatorial Africa serves as an example. Since aerial photography was the only data source that could provide a spatial resolution better than 10 m until the 1990s, it has been used for change detection for a very long period. Operations using aerial remote sensing provide the flexibility needed to minimize many issues. The workhorses of remote sensing are electro-optical scanners, which provide calibrated digital imaging in the visible and near-infrared areas of the electromagnetic spectrum. They can be thought of as being similar to digital aerial photography, although a typical instrument usually records data in several spectral bands. They cover the same spectral region as aerial photography and therefore are subject to the same limitations of inoperability through clouds or at night. Scanners have several advantages over photography, in addition to the possibility of producing multispectral data. The data are typically calibrated radiometrically, which means that a given output corresponds to a known at-satellite radiance. The data are digital, so they can be processed instantly by computers and the data can be transmitted via radio signal from the instrument to an appropriate receiving station. Although there are additional airborne devices available, this final element is particularly crucial for spaceborne data collection.

Visible and near-infrared scanner imaging has diverse applications. By using the spatial resolution, you can distinguish between them. Low-resolution (LR) satellite imagery has a spatial resolution of around one kilometer. These include coarse-resolution imagers like MISR (Multiangle Imaging Spectroradiometer), MODIS (Moderate Resolution Imaging Spectroradiometer), and AVHRR (Advanced Very High-Resolution Radiometer), which are carried on satellites in low Earth orbits, and geostationary imagers like METEOSAT and GOES (Geostationary Operational Environmental Satellites), whose main goal is to collect synoptic meteorological data(Borgne et al., 2004). The primary benefit of using coarse resolution imagers is their extensive coverage. A single swath can span several thousand kilometers up to, or in the case of geostationary imagers, the whole surface of the Earth. Thus, these imagers effectively offer very high temporal resolution at the cost/expense of coarse spatial resolution since they can swiftly gather data from large areas. Because MISR employs nine separate sensors to look at the ground, it also offers an interesting multangular perspective of the same land surface. Studies of vegetation as well as glacier studies have made considerable use of this new type of data. Additionally, the LR data are frequently made available for little to no cost. Larger glaciers, ice sheets, and ice shelves are better studied using optical low-resolution data than narrow alpine glaciers.

8.7 SATELLITE IMAGES

Based on their spatial resolution, scanners/sensors fall into the 10 to 100 m resolution range. We might refer to this type of imagery as high-resolution (HR) satellite imaging. The sensors on board the Landsat series of satellites serve as a good representation of this most common class of spaceborne scanners. The Multispectral Scanner (MSS), which had four spectral bands and a spatial resolution of 80 m, was carried by the earlier Landsat satellites until 1993. The later satellites, which began to fly in 1982, carried the Thematic Mapper (TM), which had six spectral bands in the visible, near infrared, and SWIR and a spatial resolution of 30 m. The 80 m resolution of MSS was too coarse for alpine glaciers, therefore the enhancement in spatial resolution was a significant step forward for glaciological studies. With a spatial resolution of 15–30 m, the Enhanced Thematic Mapper (ETM+), which has been using Landsat 7 since 1999, features six spectral bands in this region of the spectrum. Although the cost model has changed over time and Landsat data, including the historical archive, is now free, Landsat photos have historically cost a few hundred euros. Data of this kind have found widespread application in glaciology, for several reasons, including the good match between the spatial resolution and coverage of scanner images and the requirements for examining glaciers. While ASTER (Advanced Spaceborne Thermal Emission and Reflection Radiometer) has replaced Landsat ETM+, which has limited functionality since 2003, it has a similar temporal resolution and a spatial resolution of 15 m in the VNIR area and 30 m in the thermal infrared. Even though Landsat TM/ETM+ has a higher spatial resolution than SPOT HRV (Satellite Pour l'Observation de la Terre High Resolution Visible) and SPOT 4, their data cannot be used for cloud detection since SPOT HRV (1, 2, 3) lacks SWIR bands. Even while many glaciers fit inside the 60 by 60 km range, SPOT's aerial coverage renders it relatively inappropriate (Cawse-Nicholson et al., 2021).

Poor temporal resolution of 16 days for Landsat and 26 days for SPOT is the primary drawback of using Landsat and SPOT imagery for glacier research. A 10-day revisit period is feasible in the Alps because of the overlapping pathways (shorter at higher latitudes), but in reality, there are far fewer possibilities to obtain surface images due to frequent cloud cover. If you need to capture an image at a specific time, like the end of the glacier year when there is no more snow on the surface, this can present a significant challenge. Researchers presented limitations imposed by cloud cover on multitemporal visible satellite data in the European Arctic.

They discovered that only 7 to 15% of the images are useful in the Russian High Arctic and Iceland, and only 15 to 27% in Svalbard, Norway, and Novaja Zemlja (Russia). The high probability of getting cloud-free footage/images, ranging from 38 to 54%, was found in East Greenland. Consequently, to guarantee that there exists remote sensing data for the critical time window, an aerial photo acquisition is frequently ordered. Based once more on spatial resolution, the primary class of scanners has resolutions of a few meters down to one meter or even finer.

A commercial satellite called QuickBird, whose sensor offers digital panchromatic (black and white) images with a resolution of 0.6–0.7 m and 4-band multispectral imaging with a resolution of 2.4–2.8 m, serves as an example of such an instrument.

As a result, the swath width of 16.5 km is relatively narrow. The characteristics are comparable to those of IKONOS, another source of VHR data. Thus, despite often having lower spatial resolution, this type of imagery serves as a replacement for aerial photography. Similar uses were demonstrated while using IKONOS stereopairs over Hintereisferner, Austria, and Engabreen, Norway, to map glacier margins and create orthoimagery and elevation models. Occasionally, VHR imagers can be configured to capture stereo imagery, which facilitates the creation of digital elevation models and three-dimensional visualizations above glaciers. The drawback of all the VHR data is its relatively high cost, which might reach several thousand euros or higher if the buyer orders the image acquisition. Archived photos are a lot less expensive.

Visible/near-infrared multispectral scanners are frequently used in conjunction with scanners intended to assess the brightness temperature of the radiation in the thermal infrared. For instance, a band spanning from 10.4 to 12.5 μm is included in both the TM and ETM+ data delivered by Landsat satellites. Because of their much longer wavelength and low signal, thermal infrared bands often have a poor spatial resolution than their corresponding visible/near infrared bands. Thermal infrared scanners can be used at night since they identify radiation that is emitted rather than reflected from the sun. They are unable to pass through cloud cover, though.

Figure 8.1 LiDAR

Source: https://commons.wikimedia.org/wiki/File:RMR-Lidar.jpg, Creative Commons License

8.8 LIDAR

Short pulses of EMR are transmitted from a laser via laser scanners, also known as LiDAR (Figure 8.1) or laser profilers. Although the wavelength range is usually in the near-infrared, such as 1064 nm, bathymetric applications have also used LiDARs operating in green light (Kinzel et al., 2013). When the laser pulse comes into contact with a solid or liquid interface, it is reflected from it as long as it is not absorbed by it. The laser pulse travels through the atmosphere at a speed very near to that of light. A portion of the green laser pulse also passes through the shallow water column in this instance, reflecting off the lake or sea bed. If the position and orientation of the instrument are known, the location of the reflecting point can be ascertained through measurement of the two-way propagation time, which also provides the range from the instrument to the reflecting surface. To create a dense two-dimensional pattern of reflecting points on the surface of the Earth, modern aerial laser scanner systems send out tens or hundreds of thousands of pulses per second. Precise GPS techniques are used to establish their orientation and position. Laser scanners still have a pretty slow data rate, even though It is rather high for an aerial system. Gathering data from a typical small glacier, like Hintereisferner, which has an area of about 10 km2, takes about 30 minutes. Thus, measuring the topography of sea beds, structures, glaciers, and forest canopies is a perfect use for laser scanners. In addition to producing accurate digital elevation models (DEMs), an aerial laser scanner can achieve point densities of one or more points per square meter and vertical accuracy as high as 5 cm. This density and precision are sufficient to provide many glaciology applications.

Satellite instruments exist, however airborne sensors provide almost all of the data acquired from laser scanners. Launched in 2003, the ICESat spacecraft carries the Geoscience Laser Altimeter System (GLAS) (Wang et al., 2011). With an along-track spacing of about 180 meters, it has a spatial density of reflecting points substantially lower than aerial systems, although having a similar vertical precision. It is not well suited for studying all but the largest glaciers because of this. Ground-penetrating radar and laser scanning are similar in that the device transmits brief pulses of radio wave radiation, typically between one and twenty meters, and then detects and analyzes the return signal. Because of refraction, the wavelength is much shorter in the ice. When freshwater ice is cold, it becomes mostly transparent at these wavelengths, and the pulse travels through the ice and is reflected off the glacier base. There are additional reflections from the ice stratification. Thus, in a glaciological context, ground-penetrating radar is primarily used to examine a glacier's internal structure. Given that it is not often an airborne or even spaceborne approach, it may not fully fit the description of remote sensing. The radar equipment is typically dragged across the glacier surface by a snow scooter or other comparable vehicle.

8.9 SAR

An active imaging method that operates in the microwave or MW region of the electromagnetic spectrum is called Synthetic Aperture Radar (SAR). It operates through cloud cover and is not dependent on sunlight. Through the use of an antenna, the SAR instrument sends microwave radiation towards the Earth's surface. The radiation that the Earth's surface reflects or scatters is then utilized to create a two-dimensional representation of the variance in the backscattering coefficient over the scene. Any imaging radar would actually fit under this simplified description; however, what sets a SAR apart from other types is that it uses the forward motion of the antenna, carried out on an airborne or spaceborne platform to mimic the effect of an extremely long antenna and, as a result, achieve a high spatial resolution in the forward (azimuth) direction. It is possible to obtain high resolution in the sideways (range) direction by sending brief radiation pulses and time-resolving the return signal. The antenna must so point sideways as opposed to downward, vertically. The signal processing used after the data are gathered has a major role in determining the spatial resolution of a SAR, which can vary from 5 to 1000 m for a typical spaceborne system. Similar to visible/near-infrared systems, small swaths are typically linked to fine spatial resolution. For instance, there are several modes of operation available for the RADARSAT device. On the other end of the spectrum, it can be used in a mode with a spatial resolution of 100 m and a swath of 500 km. The best spatial resolution is 10 m with a 45 km swath width.

While SAR and visible/near-infrared images have many commonalities, they also differ significantly. The resulting image is similar to a black-and-white photo because the first-generation SAR systems, ERS-1/2 and Radarsat-1, only gave a single channel of data, or a single wavelength and polarization (Dwivedi, 2018). However, the SAR image is extremely noisy when compared to a photo. This results from interference between the radiation reflected from nearby regions of the Earth's surface since the dispersed and transmitted radiation is coherent, or has a clearly defined phase. This condition, known as speckle, lowers the radiometric resolution. Averaging can help lower it, but doing so will compromise the spatial resolution. SAR imagery must always be obtained from an oblique viewing angle. If the region being photographed exhibits significant variations in altitude, it might have multiple implications on the appearance of a SAR image. Foreshortening is a distortion caused by topography in which the radar's ground track is affected by terrain while viewing an elevated location. As a result, images of slopes facing the radar are foreshortened. The relief of the mountain is so great that, before the energy pulse from the antenna reaches the mountain base, the peak backscatters energy toward the antenna. This phenomenon occurs when a material has a lower backscattering coefficient. Ultimately, no return signal will be received from the other section of the terrain if one portion of the terrain absorbs the microwave radiation from the radar and keeps it from reaching that other portion. This is known as shadowing, but unlike optical shadows, radar shadows don't contain any information about the ground surface because the radar signal never gets there.

8.10 ADVANTAGES OF SAR

One of SAR's main advantages over VNIR systems is that it makes use of coherent radiation. Since the wavelength of a typical SAR system is only a few centimeters, this means that a change in the relative position of the radar and a scattering object of the order of one wavelength can produce a measurable change in the scattered signal from the object. This could lead to extremely precise geometric measurements. The foundation of interferometric SAR (InSAR) is this. With InSAR, one may essentially do two kinds of measurements. The first of them involves using SAR images from two or more marginally different places to determine the three-dimensional geometry of static objects. Using this, DEMs over glaciers can be created. The second kind of measurement uses SAR images taken at two separate periods and from almost the same position to calculate an object's velocity. This has been employed to gauge the glacier flow. Implementing InSAR is not a very simple task. Repeat-orbit interferometry, in which data is obtained from the same satellite many times as its orbit transports it to almost the same position, is frequently used to achieve it. The ENVISAT satellite SAR is an example of a spaceborne system. This has a wavelength of 5.6 cm for both HH and VV polarizations, and it may be used in various modes to provide swath widths of around 60 to 400 km and spatial resolutions of 30, 100, or 1000 m. It is possible to detect many glacier parameters using terrestrial, aerial, or spaceborne remote sensing data, including albedo, reflectance, surface temperature, melting, glacier zones, glacier area, equilibrium line, mass balance, glacier surface and bed topography, glacier volume, and glacier velocity. In many situations, remote sensing is the only practical option for detection (glacier areas), but it can also be used as a substitute for labor-intensive glaciological field work. Remote sensing is the technique that can offer information on glacier changes from every continent as well as from a significant number of glaciers and ice sheets, allowing glaciers and their variations to be used as indicators of climate change or as an early warning system for sea level rise. Several programs, such as Global Land Ice Measurements from Space (GLIMS) and the World Glacier Monitoring Service (WGMS), have gathered and disseminated glacier data in the past and present (Pan, 2013). On the other hand, although present and many future instruments may facilitate the job and generate new applications and possibilities for discovery, remote sensing may never completely replace glaciological field work and observations, which are always essential as ground truth data.

8.11 SNOW ACCUMULATION MEASUREMENTS

Snow accumulation that does not melt during the summer months gives rise to glaciers. The buildup of snow turns into glacier ice, which moves under its own weight when it occurs year after year. Wet and cold climates are ideal for the formation of glaciers because of the snowfall and snow buildup over several years. Monitoring can be done through on-the-ground/in situ measurements using different glaciological, hydrological, or geodetic techniques; however, satellite and aerial remote sensing offer more accurate, affordable, and practical data and monitoring techniques. There are three possible ways for solar energy to reach the glacier's surface: reflection, absorption, or transmission. To map glaciers, the primary property in optical remote sensing that is measured and utilized is reflectance. Every glacier zone has a different reflectance characteristic, including firn, dry snow, wet snow, and blue and white ice. Furthermore, differences in the spectral fingerprints of different surface types are brought about by surface aging, varying amounts of contaminants, and debris cover on the glacier surface. According to spectral reflectance analysis, snow has a high reflectivity in the visible range and a significant drop in the intermediate and shortwave infrared and near-infrared regions. Aging contaminated snow is an anisotropic reflector with a substantial specular component, in contrast to freshly fallen dry snow, which behaves as a Lambertian reflector with upward fluxes that are almost equal in all directions. Forward scattering behavior is seen on glacier surfaces, with increased anisotropy from dirty surfaces(Knap & Reijmer, 1998).

Snow that has just fallen has the highest reflectivity in the visible and near-infrared wavelength range. The reflectance of glacier ice decreases rapidly in the red (600–700 nm) wavelength region, although it stays quite high in the blue (400–500 nm) and green (500–600 nm) wavelength bands. On glacier ice, dirt (dust, volcanic ash, debris) significantly reduces reflection, and there is no discernible fluctuation in reflectance. However, contamination of snow, firn, and ice reduces the reflectance by up to 40% for ice, 10 to 80% for snow, and 10 to 50% for firn. In the near-infrared spectrum, contaminated ice appears to reflect light better than clean ice. Furthermore, for polluted surfaces, the difference in reflectance in each wavelength band appears to be substantially smaller.

8.12 ALBEDO

Albedo, which ranges from 1 to 0, is the ratio of radiation incident on a surface to radiation reflected from that surface. Albedo is a ratio of all incident and reflected solar radiation in the electromagnetic spectrum, not just in the visible range. According to the proportions of electromagnetic radiation from the sun, 41% of solar radiation is in the visible wavelengths, 9% in the ultraviolet, and 50% in the near-infrared. An albedo of 0.80–0.97 is associated with a highly reflective surface, like dry snow, while 0.15–0.25 is associated with a dark surface, like dirty ice. In some areas of the electromagnetic spectrum, albedo may be a more useful metric than spectral reflectance for a variety of applications.

The transmittance and absorption increase with decreasing albedo; however, the wavelength determines the relative amounts of reflectance, absorption, and transmittance. As more energy is absorbed, the low albedo accelerates the melting of the glacier, particularly in the near-infrared wavelength range. The ice in crevasses and ice caves appears turquoise because blue and green light are transmitted more by the glacier surface than red light, which is absorbed. The albedo reduces with increasing grain size because there is less possibility that photons will be dispersed from the snowpack, particularly in the near-infrared wavelength range. The contaminants reduce albedo in visible wavelengths but less so in the near-infrared range. One key feature employed in the mapping of glacier zones using optical remote sensing data is the reflectance of the various types of glacier surfaces. The variations in reflectance, expressed as brightness values or digital numbers and converted to reflectance (%) in different wavelength bands, help distinguish glaciers from their surroundings and categorize them into different types. Because heat seeps into the ice, the thermally active surface layer of glaciers, where seasonal fluctuations affect, is located at a depth of roughly 10 m. The majority of a glacier's ablation occurs as melting. However, ablation can also occur as sublimation, which is typical for tropical glaciers. Thermal remote sensing techniques can be used to identify the wet snow zone because the presence of liquid water on the ice surface enhances absorption in red and near-infrared wavelengths. Apart from liquid water, the glacier surface may experience notable temperature variations due to debris and pollutants. As the debris warms up in the sun, the glacier melting process could accelerate if the layer thickness is no more than two to three centimeters. More insulation from a thicker layer will result in unique topographic features on the glacier.

8.13 IMAGING APPROACHES

Using the thermal infrared bands of AVHRR or MODIS for coarse scales or band 6 of Landsat TM/ETM+ for finer scales, thermal infrared imaging can be utilized to determine the temperature(T. He et al., 2018). The data has to be properly adjusted for atmospheric factors and cloud cover. The process of creating a glacier mask usually involves determining the threshold value that separates the glacier from the surrounding terrain. Applying data from laser scanners is a relatively new and promising method for evaluating surface zones. In addition to retrieving an intensity signal from the surface and the transit time needed to determine elevation, modern laser scanners may also map surface features. The area of the glacier can be calculated using a variety of techniques after the optical remote sensing data has been orthorectified. These include manual delineation of the glacier perimeter, band ratio techniques, supervised or unsupervised pixel-based classification algorithms, and object-oriented classification. Nevertheless, cloud cover, snow cover on the glacier margins and surrounding area, and debris cover on the glacier surface and surrounding area all add uncertainty to the mapping of the glacier area. Many techniques have been devised to overcome these ambiguities. Using different indices and threshold values computed from the satellite data, different masks can be generated in automated and semiautomated glacier delineation processes. Elevation models can be used to show the dead ice zones and debris-covered areas from valley slopes and flowing glaciers. This method is based on analyzing the shift in the microtopography between the glacier covered in debris and the slope, or the change in the slope angle. While optical remote sensing data has been utilized more frequently for mapping glacier areas than SAR data, many features can be identified using various bands and methods.

Wet snow and ice-free surfaces can be mapped using C-band imaging, such as that from RADARSAT and ERS-1/2, however the ability to distinguish between snow, ice, and bare rock is limited. It may be possible to increase the differentiation of rocks using texture. Snow may be distinguished from other surfaces using L-band images, especially once incidence angle fluctuations have been taken into account. Optical satellite imagery may increase the utility of SAR data. Very high-resolution digital elevation models (DEM), which may be created using aerial laser scanner data can also be used to map the area covered by glaciers. In order to map the glaciers in South Tyrol, Italy, an automated approach for delineating glaciers on a digital elevation model (DEM) was generated from laser scanner data. Except for glaciers that were extensively covered in debris, the procedure was effective. Using multitemporal datasets, the elevation changes on the glacier due to ice movement or melting were assessed in order to identify areas covered in debris.

CHAPTER 9

GLACIERS OF THE HIMALAYAS

Contents

9.1 INTRODUCTION

Globally, a large number of glaciers and snowpacks are retreating. The cause of this retreat is the rate of glacier ablation, or melting and sublimation, which is higher than the rate at which glaciers grow and accumulate over time due to precipitation. Different regions, and locations experience glacial retreats at different rates and times. The reasons for this kind of retreat are complicated, and while climate change is a significant contributing factor, It is crucial to recollect that most of the time there are several factors at play that interact in unpredictable ways. A glacier will eventually reach a point of diminishing volumes of meltwater, which will eventually become zero when the glacier disappears. There are doubts regarding what will happen to the available water supply whenever glaciers continue to recede. The water towers are mountains, and in many areas, the amount and timing of streamflow from glaciers and snowmelt are essential for hydropower production, agricultural output, water supply, and the functionality of ecosystems. The huge populations of China and South Asia depend on the main rivers in the Hindu Kush-Himalayan (HKH) region for both water and electrical generation (Rasul, 2014). The HKK region has many basins that are water stressed, and a significant portion of the predicted population growth is expected to exacerbate this stress. This has given rise to uncertainties about possible adverse effects from changes in the accessibility of water supplies in the following decades. While many glaciers have receded over the past few decades worldwide, the rates of retreat and mass loss can differ greatly, suggesting that glaciers are influenced by a range of regional-scale factors in addition to warming, such as variations in cloud cover, precipitation, aerosol concentration, glacier geomorphology, tectonic activity, and debris. Different patterns of the glacial retreat were discovered across the HKH region in recent studies on glaciers using in situ and remote sensing data. Thus, a wide range of variables influence when and how much streamflow fed by glaciers occurs, which could have a substantial impact on the availability of water and the stability of the surrounding area.

9.2 HINDU KUSH-HIMALAYAN (HKH) CONCERNS

The general consensus in the HKH region is that naturally occurring run-off from precipitation and snowmelt is significantly enhanced by glacial meltwater. Although accurate, the watersheds have complicated hydrology, making it unclear how much glacial meltwater adds to the overall amount of water in these rivers. It is uncertain exactly what effects faster rates of glacial waste will have on populations downstream. It is unclear how climate change may impact the timing, nature, and rates of glacial waste in various sections of the HKH region. Glaciers in the eastern and central portions of this region are receding at rates that are equivalent to those in other regions of the world, but in other areas of the western Himalayas, they are comparatively stable and may even be advancing. Concerns exist regarding the long-term effects of these modifications on the availability of water for residential, commercial, and agricultural purposes as well as on the different ecological applications. The Indus, Ganges/Brahmaputra, Mekong, Yangtze, and Yellow Rivers are only a few of the major river systems in Asia that have glaciers near their headwaters (Su et al., 2016). About 1.5 billion people rely on these rivers for their agriculture and drinking water supply. They also sustain essential ecological and cultural amenities and services, as well as producing substantial amounts of hydroelectric electricity. Given that Asia is one of the water-stressed regions of the world, the surface water of these rivers and the groundwater they are connected to represent an important strategic resource. About three-quarters of the region's glaciated surface is contained in the combined basins of the Indus, Ganges, and Brahmaputra.

9.3 BRAHMAPUTRA

Glaciers make up a relatively minor portion of the Mekong's water resources, despite the river's origins on the Tibetan Plateau and its significance as a water source for the people of Cambodia, Laos, Thailand, and Vietnam. Large areas of China are supplied with water by the Yangtze and Yellow Rivers, which have comparatively less glacial cover. Since there isn't much glacier land in the basins of the Mekong, Yangtze, and Yellow Rivers, precipitation and snowmelt are what drive these rivers to flow/discharge. Precipitation contributes significantly more to most basins than snowmelt. These three rivers are found in the eastern HKH region, where the summer monsoon dominates the yearly precipitation. These factors suggest that the monsoons have a larger influence on the flow in the Mekong, Yangtze, and Yellow Rivers than in the past, and the effects of climate change on the monsoon will outweigh any changes in the glaciers. The term Brahmaputra is the official name of the river once it enters India, however, the term Tsangpo is commonly used to refer to the upstream section of the river on the Tibetan Plateau. The Ganges and Brahmaputra are considered a single watershed in this discussion of concerns related to watersheds since they eventually combine before flowing into the Bay of Bengal. Geographically speaking, the HKH region is large and complex, with dynamic and potentially changing hydrological and climatological features. This indicates that because of the high spatial heterogeneity, it is exceedingly challenging to generalize observations over the whole region. The general trend of global temperatures rising has accelerated in recent years, but the precise effects of climate change in the HKU region remain uncertain due to regional variability and a lack of local data. This includes the potential rate of glacier retreat on a regional scale and the possible effects on the HKH hydrologic system.

9.4 GLACIERS OF THE HKH

The density of the snowpack steadily rises with time as a result of metamorphic processes and compaction of the snow layer above. Any snow that doesn't melt gets carried over to the following season and may get buried by more precipitation. Firn or névé refers to snow that has been around for more than a year but is not yet glacial ice. The rate at which seasonal snow turns into glacier ice is determined by the local climate. The movement of glaciers is attributed to many gravitational mechanisms, such as internal deformation brought on by the shear stress exerted by snow and ice on top of them, and basal sliding on a layer of liquid. The pressure created by snow and ice on top of one another causes ice masses to flow across slopes. A mass of ice is deemed to be a glacier if it begins to flow as a solid. Glaciers are not made up of stationary patches of snow and ice. Understanding that glaciers contain both an accumulation zone, where the volume of the glacier increases, and an ablation zone, where volume is lost, makes it easy to comprehend the basic principles of glacial activity. A glacier gains substance during the accumulation season, which is summer in the eastern Himalayas and winter in the western Himalayas. Some or all of that accumulation is lost during the melting season, which occurs in the summer in both the eastern and western HKH. Therefore, a glacier's size can fluctuate, grow, or stay the same throughout the year. The zone at higher elevation where there is an annual net gain in mass is known as the accumulation zone, and the zone at lower elevation where there is an annual net loss in mass is known as the ablation zone. The height at which the accumulation and ablation zones converge and the annual net mass balance is zero is known as the equilibrium line altitude (ELA) (Benn et al., 2005). The net difference between accumulation and ablation is the annual mass balance.

9.5 MECHANISMS

Snowfall, condensation, refreezing of precipitation, avalanche transport onto the glacier, and blowing snow transport onto the glacier are all mechanisms that contribute to the accumulation of snow and ice mass on glaciers. Ablation refers to all of the mechanisms that cause glaciers to lose snow and ice mass, including calving, melting ice, sublimation, and snowmelt. Glaciers and lakes are comparable in terms of their ability to discharge water. The total amount of water held in a lake is comparable to the water stored in glaciers. The process of precipitation and water entering a lake through streams, rivers, and groundwater channels are examples of processes similar to glacial accumulation. The process of removing water from a lake by evaporation, groundwater extraction by humans, and water carried out of the lake by rivers, and other channels is comparable to glacial ablation. The lake is in a steady condition and its level stays constant when its input and output of water are equal. The amount of water held in glaciers remains constant when accumulation and ablation are equal, and the ELA remains stationary. When ablation consistently outpaces accumulation, the ELA shifts upward, and the glacier in question eventually disappears, and glacial volumes decline. Whether the net change is positive or negative depends on the change in the glacier volume rather than its area. However, as it is challenging to quantify a glacier volume directly, there are few glacial volume data available worldwide. There won't be any accumulation and the glacier will eventually disappear if it is completely below the equilibrium line. The data on glacial mass balance will offer a vital connection between fluctuations in glacial volume and changes in the climate. To comprehend what happens to glaciers throughout time, one must have a general understanding of glacial mass balance. Glaciers adapt to the climate to reach a steady state, which is characterized by a constant mass balance and ELA over time. When temperatures drop or when precipitation increases, a glacier advances and there is a positive mass balance. A negative mass balance is the outcome of rising temperatures or falling precipitation, which causes glacial retreat.

Figure 9.1 The HKH region

Source: https://commons.wikimedia.org/wiki/File:Hindu_Kush_Mountains%27_view_,_Pakistan.jpg

Melt occurs in the ablation zones of glaciers in mid-latitude mountain ranges worldwide, including those in the HKH region (Figure 9.1), at some locations every year. The majority of mid-latitude glaciers, if not all of them, supply streams and rivers with meltwater. This glacier meltwater contribution to the outflow of rivers and mountain streams happens even in years when the glacial mass balance is positive. When the climate is such that accumulation and glacial melt are equal and the mass balance remains constant for a certain amount of time, this is known as a steady-state condition. The hydrologic cycle can be used to analyze the contribution of the glacier to streamflow. Sublimation is an important glaciological factor when assessing the mass balance of the glacier at the high elevations of the HKH, as well as in central and northern Tibet, where conditions are extremely dry and cold (Azam et al., 2021). However, hydrological considerations do not take this into account. The annual glacial mass balance is equal to the annual snowfall less the annual glacial ice melt and the yearly snowmelt from the glacier. The glacial ice melt is defined as the volume of ice melt that is equal to, or less than, the water equivalent of snow that accumulates into the glacier system in a year if the glacier is in equilibrium or has a positive mass balance. The amount of ice melt that surpasses the water equivalent of the volume of snow buildup into the glacier system, resulting in an annual net loss of glacier volume, is known as glacier waste if the mass balance of the glaciers is negative. To put it briefly, glacial melt does not automatically indicate a negative mass balance or waste. The word "glacier melt" refers to storage rather than an increase in the yearly streamflow. Glaciers and groundwater are storage reservoirs in the hydrologic cycle. Glacial melt is analogous to groundwater extraction or outflow from artesian aquifers, snowfall on glaciers is analogous to groundwater recharge, and glacial waste is analogous to groundwater overdraft.

9.6 MELT

The timing of glacial melt is significant because it stores water as snow accumulation in the glacier system, delaying water equivalent runoff until ice melts in the late summer months of the normally low streamflow. Glacial melt can have a seasonal impact on overall streamflow. As a result, seasonal timescales determine the majority of the glacier melt contribution to streamflow. Understanding ice dynamics is necessary to comprehend how glaciers are responding to climate change. There is a lag of many decades or more between a change in the climate and the advance or retreat of glaciers. Variations in the glacier terminal from year to year are probably a result of climate events that happened several decades ago. A glacier's area and volume, precipitation patterns, debris cover, and topography affect how quickly it responds. All of these variables differ greatly throughout Asia's High Mountains and HKH. One can find the terminal of a glacier by simply walking uphill to the glacier's edge. There exist precise records of the terminus position of certain glaciers that date back over a century. But over brief time spans of a decade or so, this simple measurement could provide inaccurate information/data regarding the rate and direction of a glacier retreat. When the terminal of a glacier retreats for multiple decades, it is a sign that the glacier is receding. Using a network of pits on the glacier surface, the glaciologic method of calculating glacier mass balance measures changes in surface level between two predetermined dates (annual mass balance) or at the end of the ablation and accumulation seasons (seasonal mass balance). This approach is thought to be the most accurate and yields the most data regarding geographic variation. Nevertheless, there are currently very few measurements of glacial mass balance and no long-term glaciologic mass balance records for the HKH region (Mehta et al., 2021).

9.7 DEM

One can estimate mass balance by applying the geodetic method. Using a variety of digital elevation models (DEMs) created throughout the whole glacier surface, this indirect method measures changes in the glacial surface elevation over time. Only decadal or longer time scales can be utilized to assess glacier changes using the geodetic approach due to significant uncertainty. The remote position of glaciers and the harsh topography present logistical challenges, which is why remote sensing techniques are particularly interesting. Regular monitoring of the area, length, surface elevation, surface flow, accumulation/ablation rates, and albedo of glaciers is made possible by remote sensing. The major ice fields outside of the polar regions are found in the HKH region, which is frequently referred to as the third pole. Furthermore, the Himalayan mountains and glaciers hold cultural significance for the people residing in the region. Although comprehensive surveys of glaciers are still lacking, there is consensus regarding the extent of glaciers in the area. It is estimated that there are more than 110,000 km2 of glaciers covering the HKH, the Tibetan Plateau, and the Tien Shan region, with roughly 50,000 of those glaciers being recognized. Comparing glacier areas is challenging, though, as spatial extents are frequently inconsistent or poorly classified. The mass loss rate in the area may be closer to 4 ± 20 Gt yr-1, despite earlier estimates ranging from 47 to 55 Gt yr-1. The disparities and gaps in these different assessments highlight the necessity of doing a thorough glacial inventory of the area. Furthermore, a better knowledge of the amount of glacial area in susceptible low-elevation locations i.e., below the ELA would result from additional information regarding the distribution of glacier area with elevation. Plotting the distribution of glacier area with elevation is known as glacial hypsometry. A glacial hypsometry for Nepal has been created. The hypsometry reveals that the elevation of the glacial ice varied from roughly 3,200 m to 8,500 m. The 100-meter elevation zone, which was centered at 5,400 meters, has the greatest quantity of glaciated terrain.

Precise measurements of the glacier area are not available. Researchers compute the percentage of glacier area for the Indus and Ganges/Brahmaputra in various elevation bands. The bulk of the glacier area in both basins is located in the 5,000–6,000 m band, with a sizable portion in the 4,000–5,000 m band. While the Ganges/Brahmaputra Basin has a slightly bigger proportion of its glacier area above 6,000 m, the Indus Basin has a somewhat greater proportion of its glacier area below 4,000 m. These values are comparable with the more rigorous hypsometry data from Nepal, even if they should be regarded as qualitative. Although the differences are negligible, they imply that glacial retreat in the Indus basin would be more susceptible to climatic variations than in the Ganges/Brahmaputra basin; this qualifies as evidence of greater glacier stability in the western Himalayas. This is because variables other than elevation, like local temperatures, debris cover, and precipitation regime, can affect glacial retreat. Due to a dearth of field data, rates of glacier retreat in the HKH are poorly understood, which makes it challenging to assess the effects of regional climate change. It has often been demonstrated that AX010 in Nepal, one of the most researched glaciers in the area, has a negative mass balance. It is expected that AX010 will disappear by 2060. Glacier AX010 is not a reliable representative of regional patterns in the HKH (Li et al., 2016). These findings further suggest that smaller glaciers cannot be used to identify regional trends.

9.8 WATER LOSS

The rate of water loss from glacier retreat between 2002 and 2006 was estimated to be -55 Gt yr-1 for the entire region of Asia based on extrapolation of these few mass balance studies, with -29 Gt yr-1 across the eastern Himalayas alone. In contrast, researchers estimated a mass loss of just -4 Gt yr-1 for the region of High-Mountain Asia (HMA) from 2003 to 2010 using new data from the Gravity Recovery and Climate Experiment (GRACE) project. The analysis of the GRACE data yielded a much lower estimate of glacier loss. This is at least partially because the GRACE satellite information integrates over the entire region. Many studies have calculated the rate of retreat of individual glaciers in the area: the Dokriani glacier in Uttarakhand, India retreated 550 meters annually between 1962 and 1995; the Parvati glacier in Himachal Pradesh, India retreated 578 meters annually between 1990 and 2001; and the Bhagirath Kharak glacier in Uttarakhand, India retreated 7 meters annually between 1962 and 2005. The Satopanth glacier in Uttarakhand, India retreated 23 meters annually between 1962 and 2005. In general, the glaciers of the eastern HKH region have a negative mass balance and are retreating, but not more quickly than other mid-latitude glaciers. However, there is still uncertainty regarding the retreat rates of specific glaciers, such as the Gangotri glacier in northern India, and more mass balance measurements are required.

In contrast to the eastern HKH, the larger glaciers in the Karakoram region, especially those at higher altitudes, have been expanding since around 1990 in the western HKH. In the Karakoram region, 58% of the glacier fronts were either stable or advancing slowly, with a mean rate of around +8 ± 12 m yr-1. The Karakoram glaciers have been shown to surge, although additional fieldwork is required to verify whether this suggests a positive mass balance. Data from the late 1980s suggested that the Siachen Glacier may be trending toward a negative mass balance. However, more recent data from remote sensing indicates that glaciers in the central Karakoram had a marginally positive mass balance between 1999 and 2008. In comparison to the eastern part of the HKH, the western end tends to exhibit slower rates of retreat, less generation of pro-glacier lakes linked to flood hazards, and more frequent observations of advancing glaciers. The total amount of glacial ice lost in the last ten years in the region is substantially less than previously estimated. Studies on how HKH glaciers react to climate change are currently scarce. Some of the larger glaciers in the Karakoram region of the HKH appear to have positive mass balances, suggesting that the glacial mass balances there are stable. These findings imply that during the course of the next few decades, the glacier extent in this area of the HKH region will not change much. Glaciers in the eastern Himalaya are receding at a pace comparable to global averages. Glacier retreat rates may be lower than previously thought, albeit these estimates are still subject to uncertainty. For the next two to three decades, there may not be much of a change in the rates of glacial retreat and glacier extent in the eastern Himalayas. Even if this is the case, future changes in climate forcing may cause the rate of glacial retreat to accelerate. Elevations under 5,000 meters in the eastern HKH have the highest retreat rates. This is especially dangerous for glaciers whose highest point is less than 6,000 meters. It is anticipated that these low-elevation glaciers will continue to melt rapidly. The most probable areas to be affected by these retreating glaciers are high-elevation settlements and activities that rely on glacial meltwater produced by these glaciers. The research groups are unable to predict with certainty whether there will be significant changes to the rates of glacier retreat or the size of glaciers in the HKH region over the next few decades.

The albedo range of fresh snow is approximately 0.75 to 0.95. Glacial ice, on the other hand, ranges in albedo from roughly 0.3 to 0.4 (Ming et al., 2013). When snow is removed from a glacier, the energy supply to the glacier surface increases by 200–300%, while other variables such as air temperature and cloud cover remain unchanged. Added snow nearby melts when the exposed glacier ice warms up and melts, giving the glacier additional energy flow and increasing glacial waste. The Arctic Sea ice is presently going through this albedo feedback process. Another scenario is a shift in the regional monsoon activity that lowers the quantity of snowfall in the Himalayas and is accompanied by high levels of black carbon deposition, which leads to extremely high rates of waste. The rate of glacier loss would rise if higher air temperatures were imposed as a result of the atmosphere being heated by black carbon. Reduced snowfall and exposed glacier ice due to a shift in monsoonal activity could result in increased debris rates. The albedo of glacier ice and snowpack is being reduced in the HKH region due to the large amounts of black carbon that are being entrained in the atmosphere and deposited there. The persistent snowfall that occurs during the monsoon season in the Himalayan region helps to offset the lower reflectivity caused by black carbon accumulation. Snow depositions also help to decrease the buildup of black carbon. Black carbon carried from the Indo-Gangetic Plain could build on the snow surface if fresh snow does not fall due to changes in monsoonal activity, which would accelerate the rates of waste. Furthermore, the contribution of glacier waste to summer streamflow would increase in this scenario with less monsoon precipitation and more glacier waste. Such increased rates of waste could occur even in the absence of a temperature change in the atmosphere. On the other hand, high atmospheric concentrations of black carbon warm the atmosphere, which would quicken melt rates even more. Black carbon may also have a significant impact on altering the precipitation phase, such as from snowfall to rainfall. The amount of snow days and precipitation in the area may potentially change as a result of climate change. To comprehend whether snowfall occurrences in the region are changing and how such variations will affect glacial mass balance, further data on the precipitation is required. Accelerated rates of glacial retreat beyond current rates are possible under certain circumstances.

9.9 MOUNTAINS AND CLIMATE OF THE HKH

The HKH region is home to some of the steepest mountain ranges. The range extends from its base in the alluvial Indo-Gangetic Plain to the Tibetan Plateau (8,000 m). The Himalayas, along with the Tibetan Plateau, have a significant impact on the Asian monsoon system. As a result, the region has a very strong climate gradient. At the base, the climate is tropical; at the highest points, it is permanently covered with snow and ice. The surface of the Plateau warms up rapidly in the late spring and early summer, acting as an elevated heat source that pulls warm, humid air from the Indian Ocean toward the Tibetan Plateau and the Himalayas (B. Chen et al., 2012). Over the Indo-Gangetic Plain and the Bay of Bengal, the monsoon flow causes intense monsoon rains as it carries moisture from the Arabian Sea to the Indian subcontinent. The low-level monsoon flow reverses to the northeast throughout the winter, resulting in comparatively dry weather over India. Rainfall on the Tibetan Plateau is infrequent, averaging between 100 and 300 mm. In the winter, the majority of precipitation falls as snow, covering more than 50% of the territory above 5,000 meters in altitude. At the same elevation, the percentage of snow cover decreases to less than 30% during the summer. When the snow melts, an arid landscape with sporadic glaciers and huge brackish lakes appears. The Sutlej valley roughly divides the climate regimes of the western and eastern Himalayas. The mid-latitude westerlies are responsible for around two-thirds of the high-altitude snowfall in the western Karakorum. The summer monsoon contributes almost 80% of the annual precipitation in the east. High mountains and glaciers of the Hindu Kush and Karakoram may be found in the westernmost part of the region. Many rivers flow into the upper Indus River Basin before eventually emptying into the northern Arabian Sea. This area borders the mountainous, arid areas of Afghanistan to the west and northwest Thar Desert to the south. Its climate is characterized by relative dryness, with 400 to 600 mm of precipitation falling annually, mostly from winter storms linked to mid-latitude westerlies. Although the mean annual precipitation in the cold-arid parts of Ladakh, India, is as low as 115 mm per year, the region experiences somewhat higher summertime precipitation (Wani et al., 2020).

9.10 AEROSOLS

Aerosols can originate from both natural and artificial sources. In the eastern HKH region, aerosols originating from natural sources like desert storms have long been recognized to coexist with the Indian monsoon. The mid-latitude westerlies carry dust from the Middle Eastern, Afghan, and Thar deserts in northwest India to the Indo-Gangetic Plain and the Himalayas in April and May. The Indo-Gangetic Plain is one of the most polluted areas of the globe because of the significant rise in atmospheric loading of aerosols from man-made sources like factories, power plants, cooking and heating, and agricultural practices since the Industrial Revolution. These aerosols frequently take the shape of atmospheric clouds which are brown in color. Black carbon is a major ingredient in these dark clouds. Internal combustion engines, power plants, boilers, waste incinerators, agriculture, and forest fires are examples of sources of black carbon. Although certain types of aerosols have a cooling impact, airborne black carbon heats the atmosphere by rapidly absorbing solar radiation. Aerosols, especially black carbon, which can warm the atmosphere can modify cloud and rain formation processes, change the radiation balance of the Earth's surface and atmosphere, and have an impact on regional and global water cycles, which include the Himalayan snowpack and glaciers. Aerosols play a crucial role in the monsoon climate system because they can control surface evaporation, river runoff, atmospheric heat sources and sinks, rainfall during the monsoon, and the melting of glaciers and high mountain snowpack. Aerosol optical thickness (AOT), a quantitative measure of the amount of solar radiation attenuation at the Earth's surface caused by particles, is used to quantify the atmospheric loading of aerosols. The AOT peaks over the Indo-Gangetic Plain and northwest India in April and May, the late spring and early summer seasons. The Tibetan Plateau blocks the monsoon flow, forcing it to move across northern India and the foothills of the Himalayas. Aerosols from distant deserts and local emissions consequently travel significant distances (> 5 km) against the Himalayas before dispersing throughout the Indo-Gangetic Plain and areas south of it. Furthermore, the Indo-Gangetic Plain and the Himalayan foothills receive a rising amount of warm, humid, and unpredictable oceanic winds from the Arabian Sea and the Indian Ocean due to the southwest monsoon flow. The deep aerosol layer combination of black carbon and dust heats the atmosphere effectively. It enhances the atmospheric water cycle feedback through interactions with the warm, humid monsoon air, and it has the potential to dramatically alter the summer monsoon rainfall.

9.11 EMISSIONS

Local aerosol emissions from the Indo-Gangetic Plain are trapped in the stable, low boundary layer (<1 km), which is carried by the northeasterly flow as atmospheric brown cloud plumes that radiate from the source region over the Indo-Gangetic Plain to the ocean. Because there are fewer desert winds in the winter, local black carbon emissions contribute more to atmospheric brown clouds than dust does. The boundary layer is heated by the black carbon aerosol, but the land surface is cooled. This increases atmospheric stability by convection and the already insufficient winter rainfall. It is commonly recognized that the heavy aerosol loading that occurs throughout the winter months in the boundary layer of the Indo-Gangetic Plain negatively affects vision, aircraft safety, and human health. However, because the stable boundary layer and the free atmosphere above are effectively separated from the aerosol forcing, the aerosols will have little effect on the large-scale atmospheric water cycle. Aerosols have an impact on monsoon temperature and precipitation, and variations in the South Asian monsoon climate have a significant impact on the growth and deterioration of the Himalayan snowpack and glaciers. Aerosols cool the Earth's surface below their layer by reducing surface solar radiation, a phenomenon known as the dimming effect. The Asian monsoon may decline as a result of the cooling, which also lessens the north-south sea surface temperature gradient and land-sea thermal variation. The aerosols of desert dust found over the Indo-Gangetic Plain are more absorbing than natural dusts from other regions of the world. Dust particles that were carried over the heavily industrialized areas of the Indo-Gangetic Plain may have picked up small soot particles, making them more effective solar radiation absorbers. Due to the atmospheric water cycle feedback, or the so-called Elevated Heat Pump (EHP) effect, these highly absorbing aerosols may have an important role in the anomalous heating of the atmosphere over northern India and the Tibetan Plateau (Tang et al., 2018). This phenomenon causes the monsoon rainfall to shift northward toward the Himalayan foothills and northern India during the late spring to early summer monsoon season.

Because of the wide aerosol layer over the Indo-Gangetic Plain, as well as its interaction with the warm, humid, unstable atmosphere, it is believed that the EHP effect is most effective in May and June. On the other hand, the wintertime cold, dry air mass and the confinement of aerosols within the narrow zones over the Indo-Gangetic Plain and atmospheric boundary layer reduce the EHP effect. Based on experiments using a global climate model, the EHP effect produced by dust and soot over the Indo-Gangetic Plain in late spring and early summer could cause the Himalayan seasonal snowpack to melt more quickly due to aerosol-induced atmosphere-land surface interaction. However, some studies have discussed the significance of the EHP based on observations. Many studies involving global modeling have proposed that non-local processes could account for a significant portion of the aerosol driving of monsoon variations. These include changes in radiative fluxes reaching the ocean surface that power the monsoon, remote sea surface temperature forcing, and large-scale atmospheric circulation changes. Several studies have demonstrated indications of a weakening summer monsoon over the previous fifty years. There will be less rain across India and somewhat more over the Tibetan Plateau. The timing and intensity of the South Asian monsoon can be significantly disrupted by aerosols, however further research is required to pinpoint the exact scale of these effects and weigh the relative significance of the many hypothesized causes. The snow-darkening effect is another important way that absorbing aerosols may lead to early snowpack melting and faster glacier retreat. Impurities detected in the Himalayan snowpack and glaciers have been identified as dust and soot that have been carried from emission sources in the Indo-Gangetic Plain and from remote places (Tripathee et al., 2021). These contaminants contributed to a distinctive decrease in the reflectivity of the snow and ice surface in the visible range. This resulted in a decrease in the albedo of the snow and ice, increased absorption of solar radiation, and a potential acceleration of the pace of snowmelt and glacier retreat.

9.12 DUST AND BLACK CARBON

Recent research in the HKH region suggests that dust and black carbon-induced snow darkening could be the reason behind glacier retreat. Based on observed ice core samples in Tibetan glaciers, researchers hypothesized that black carbon deposition may have played a role in the quick retreat of the glaciers. Because of their higher snow-turnover rate, large glaciers like those in the western HKH will be more impacted by black carbon deposition than small glaciers with steep slopes like those in the eastern HKH. This shows that there is some doubt about the impact of black carbon or BC on glacier retreat, but it also shows that there are other variables that affect glacial retreat and that these variables interact in unpredictable ways. Increased aerosol loading across the Indian subcontinent between 1990 and 2000 resulted in a 0.9 percent decrease in snow/ice cover over the Himalayan area. Black carbon emissions from coal and biofuels played a major role in this decline. There is a significant reduction in the Himalayan snow/ice albedo forcing when black carbon and other pollutants are reduced. Nevertheless, it is unclear that specifics of aerosol movement throughout the Himalayan ranges are represented due to the use of low-resolution models. Confirmation of the outcomes of these model investigations requires additional data. Using remote sensing, the impact of various snow cover types, including black carbon, on albedo has been studied. It is necessary to conduct surface observations to determine how much black carbon is deposited on glacier surfaces and how much this influences the albedo. Although the reported total black carbon concentrations were surprisingly large, raising some concerns about data quality, researchers found increased black carbon concentrations in an old deep ice-core consistent with black carbon being transported to high elevations of the Himalayas from anthropogenic sources. Researchers estimated a 6–8% reduction in surface albedo by comparing snow albedo before and after significant dust storms using MODIS satellite data. Due to a scarcity of instrumental records and insufficient sampling, measurements of dust and black carbon and their impacts on surface albedo are still rare and with uncertainty.

9.13 TEMPERATURE AND PRECIPITATION PATTERNS

Between 1988 and 2008, there was a rise in the daily maximum and minimum temperatures of 1.0 to 3.4 °C recorded throughout the Himalayas (Mir et al., 2015). From the late 1960s to the mid-2000s, it was discovered that the maximum daily surface temperature in Nepal was rising at a pace of 0.5 to 1.0 °C every decade. The Himalayas have a narrow gradient in temperature rise, with slower warming rates or even small cooling tendencies observed at lower elevations and faster warming rates at higher elevations. Surface warming is occurring at a rate that is more than five times faster than greenhouse gas-induced global warming, indicating the significance of local heating mechanisms. However, it should be emphasized that a large number of the high-elevation stations are not truly high-elevation because they are situated in deep valleys. They are not evenly distributed and are frequently found in cities or villages, which raises the possibility that the data would be affected by urban heat island effects. The patterns are less evident in Karakoram and the northwest Himalayas. There has been a downward trend in both the maximum and lowest temperatures, which are 1.6 and 3.0 °C, respectively. There is a general warming trend observed with increases of 0.06 °C per decade during the monsoon season and 0.14 °C per decade during the winter. There has also been evidence of a rising tendency in winter precipitation across the same area, which is most likely related to modifications in the mid-latitude westerlies. But between 1866 and 2006, there was no trend in winter precipitation and a decline in monsoon precipitation (northwest Himalayas and Karakoram). These observations are derived from in-situ historical observations from a limited network of stations. Despite offering a crucial scenario of the climate trends in the HKU area, their lack of geographical and temporal resolution, potential data inhomogeneity, and local influences make them questionable.

Critical geographical and temporal information about changes in temperature and precipitation in the past few decades in the HKH region has been revealed by satellite remote sensing data. A broad warming trend in the troposphere column above the Himalayan-Gangetic area has been detected by microwave satellite observations. Over the western Himalayas and northern Pakistan, the warming trend is most noticeable in the pre-monsoon season (April to May). Between 1979 and 2007, the greatest warming rate was almost 0.8 °C per decade. Large seasonal changes in the tropospheric warming trend were found, with higher warming over the western than the eastern Himalayas and statistically significant warming over December to May. Remarkably, in recent decades, tree ring records in Nepal have indicated a pre-monsoon tropospheric warming along with surface cooling, which is consistent with the direct effect of absorbing aerosols in warming the atmosphere and cooling the surface. With both temperature and aerosol exhibiting a consistent rising trend between 1979 and 2007, it is obvious that there is high coherence between interannual variations of tropospheric temperature and aerosol index for absorbing aerosols satellite measurements over the Himalaya foothills and the Indo-Gangetic Plain. With a single scattering albedo of 0.89 ±0.01 (550 nm), the aerosols across the Himalayas and Gangetic Plain are determined to be highly absorbing based on field observations of aerosol optical characteristics and surface solar flux. When the value of single scattering albedo is measured by the vertical distribution of aerosols from the spaceborne Cloud Aerosol Lidar with Orthogonal Polarization (CALIOP) observations, a model simulated solar heating profile with maximum heating in the middle troposphere (4 to 5 km) in the range of 2 to 5 °C per day is produced (Ohneiser et al., 2023). Early summer monsoon rainfall across northern India has increased since the 1950s, with an increase of more than 20 percent between 1950 and 2004. This increase coincides with the increases in tropospheric temperatures. These findings support the notion that a wet monsoon may eventually develop in northern India in the late spring and early summer, followed by a dry spell in central and southern India. Recent observations of increased pre-monsoon rainfall over northern India are also consistent with these results.

9.14 MODELING AND RESEARCH FINDINGS

An increasing number of recent modeling studies have proposed a causal connection between the increasing loading of absorbing aerosols in the Indo-Gangetic Plain and variations in the Himalayan snow cover. High black carbon aerosols in India are responsible for around 30% of the estimated 5–10% drop in snow cover in the Himalayas between 1990 and 2001, based on satellite data. Carbonaceous aerosols have an impact on the decrease in snow and ice cover in the spring in Eurasia that is similar to that of greenhouse gases. The influence of black carbon and dust heating the atmosphere can cause the snowpack in the western Himalayas to thaw early in April and to melt 8–10% faster in May and June. The re-distribution of rainfall in the Asian monsoon from pre-industrial to modern climatic forcing can be significantly impacted by the deposition of black carbon in snow cover across the Tibetan Plateau and the Himalayas. The Himalayan glaciers and snowpack are retreating more quickly, and the effects of greenhouse gas warming may not be the only factors contributing to regional effects of atmospheric heating and the darkening of the snow and ice surface by dust and black carbon. Black carbon may also have a delayed influence on glaciers and snowpack. Due to black carbon being flushed by snowmelt that re-freezes at lower layers, studies in Tibet's Tien Shan Glacier revealed that the content of black carbon at the bottom of the snowpack is significantly higher than at the surface. The black carbon-rich snow may be re-exposed if the ELA rises in response to rising temperatures, increasing solar absorption and speeding up the retreat of the snowpack or glacier.

Research is still being done to determine how aerosols may have impacted the Asian monsoon and the potential retreat of the Himalayan glaciers, as well as how they may do so. The tropospheric temperature gradient may be increased and atmospheric feedback may be induced, which would increase the amount of moisture and rainfall that reaches the Himalayan region in the late spring and early summer. The Himalayan snowpack will experience early and accelerated seasonal waste as a result of this influence. Through atmospheric land-surface and cloud feedback, lower land surface temperature during the early monsoon could result in an early termination or reduced peak monsoon. This is because of the increased cloudiness and rainfall during this period. The extent to which these scenarios have contributed to the explanation of past changes in monsoon rainfall is a topic of discussion and current research. The fluctuations in precipitation in the area, which are linked to the unpredictable Asian summer monsoon and the winter precipitation regime of the mid-latitude westerlies, have a significant impact on the future of the Himalayan glaciers. Improved modeling skills and more accurate data from space and ground-based observations are needed to comprehend and forecast the trends in temperature, precipitation, and glacier retreat, as well as how these factors affect river runoff and freshwater availability and the populations living downstream. A comparison of the historical simulations and observations serves as a major basis for evaluation. There has been a general increasing trend in temperature, especially during the last three decades. Even though there are fewer observation stations at higher altitudes and hence a potential bias in the data, evidence suggests that temperature is rising more quickly with altitude. Over the past 50 years, precipitation has decreased in the monsoon-dominated eastern and central Himalayan regions, but there has been no discernible trend in precipitation in the western Himalayas during the monsoon season (Kumar et al., 2021). In the western Himalayas, winter precipitation is trending upward. It is unclear how much the climate variability and different forcing factors contribute to these trends. However, one of the most crucial measures of the accuracy of model projections is still the model's capacity to replicate these trends.

9.15 FORECASTS FROM CLIMATE MODELS

Forecasts from climate models indicate a general and seasonal warming across the region, with a larger warming over the Tibetan plateau than over the South Asian continent south of the Himalayas. The historical temperature trends are often well reproduced by the models, giving rise to confidence in their predictions. However, model predictions of precipitation differ greatly from one another, and they frequently fail to adequately reflect historical patterns. The most consistent trend, observed in most models for the 21st century, is a long-term projected rise in summer precipitation in the Eastern Himalayas and Bangladesh. As a result, the models used to support the IPCC AR4 or Intergovernmental Panel on Climate Change consistently predict higher annual mean river discharge in the Ganges and Brahmaputra. While a significant portion of models do not, most indicate a drying trend in winter precipitation in the central Himalayas. Models are roughly equally likely to predict increases or declines in precipitation during either season over the western Himalayas. Many models display trends that change over time. For instance, the Himalayan foothills (sub-Himalaya or the Shivaliks) may see negative rainfall trends in the following ten years but increased precipitation in the second half of the century. Data from next-generation models used to support the IPCC AR5 were examined; these results could offer more reliable predictions of future changes in precipitation in this area. A deeper understanding of the role that black carbon aerosols play in influencing changes in precipitation in this area might also come from more research. Given that many AR4 (Fourth Assessment Report) models did not incorporate black carbon in their forecasts, this appears to be significant and may be a contributing factor to some of the variance observed amongst models. There seems to be a little increase in consistency across global and regional models on the projection of a rise in the frequency of extreme precipitation events with global warming. A shift in unpredictability like this could have more dire repercussions than a shift in the average amount of precipitation. Nevertheless, the poor capacity of models to faithfully capture observed variability, particularly in the monsoon systems, reduces confidence in these estimates.

9.16 METEOROLOGY

There are few meteorological records from the HKH region, and the ones that do exist mostly go back to the middle of the 1950s and the beginning of the 1960s. Thus, a lack of long-term meteorological and glaciological observations from the higher and more remote elevations of Asia makes it difficult to estimate climatic change there, particularly its true long-term significance and potential future impact. However, data indicates that the Tibetan Plateau and the eastern Himalayas are warming normally and that this warming is getting stronger as one ascends higher, at least to 4,800 m (You et al., 2020). The situation is more complicated in the western Himalayas where, since at least the 1980s, all but the Karakoram have warmed. The primary regional climate anomaly is still the Karakorum, which is also the case in relation to present glacier activity. Unfortunately, because the observation records are too short, these contemporary climate studies are unable to offer any relevant insights into the relative significance of the recent observed changes in climate with respect to the inherent variability of the regional climate. This gives rise to the foundation for reconstructing the climate over the HKH using ice cores and annual tree-ring chronologies spanning centuries. Although moraine dating lacks the temporal resolution of tree-chronologies, it can nevertheless yield valuable information. More specifically, it can offer proof of historical glacial activity. Studies on glacier motion have revealed both temporal and spatial variability in the regional glacier advance and retreat. The HKH glaciers have not always reacted the same way to climate forcing. For instance, researchers found that glacier advance was not synchronized during the last glacial cycle using moraine dating in the Kunlun Mountains, which are situated near the northern boundary of the Tibetan Plateau. Research has indicated that the spatial and temporal patterns of the monsoon and the glacial advance and retreat in the area may be related. Patterns of glacier advance and retreat are correlated with monsoon variability in northern India.

9.17 DATING AND TREE RING RECORDS

Using dendrochronological dating techniques to reconstruct glacier fluctuations from dead trees covered by advancing glaciers, tree-ring chronologies have been used to reconstruct climate over the HKH, but they have not yet been used to provide a direct history of glacial advance and retreat in the Himalayas. This is since the majority of the HKH glaciers lie above the treeline, which drastically limits the amount of tree-ring evidence available for timing previous glacier advances and related retreats. The Himalayan studies have deduced probable variations in past and current glacier activity from variations in tree growth assumed to be associated with climate (temperature). Therefore, options are limited to reconstructing the climate over the HKH using tree rings, assessing how anomalous recent climate changes are in comparison to historical trends, and estimating the extent to which those anomalous changes may be accountable for the observed retreats of the HKH glaciers.

In certain highly stressed locations, such as temperature change from tree growth at upper treeline locations, it is frequently still possible to infer past changes in climate from tree-ring records even when it is not possible to directly calibrate them against climate data, as can be the case in the data-scarce regions of the HKH. Nevertheless, compared to direct calibration and validation using instrumental temperature data, such conclusions are significantly weaker for interpreting previous changes in climate from tree growth. The study of past climate in the Himalayas through tree rings started more than 20 years ago and picked up speed in the 1990s(Yadav et al., 2015). Nonetheless, for large areas of the HKH and for specific seasons, the deductions on historical and present-day climate change derived from tree rings in the HKH are largely consistent. Over much of the HKH, there is evidence of recent climate warming since the LIA, with the later decades of the 20th century frequently being the warmest overall. This can be shown from calibrated or inferred tree-ring reconstructions of historical summer temperatures in the Karakoram region of northern Pakistan and the western Tien Shan region of Kirghizia. Regarding climatic change since the LIA, the temperature records derived from tree rings in the western and central Himalayas of India and Nepal are a little less seasonally consistent. Researchers reported tree-ring evidence for anomalous winter warming in the western Indian Himalayas, which has become increasingly apparent since the mid-20th century. Similar proof of long-term warming was discovered by scientists in their reconstruction of the cold season temperatures across all of Nepal. Spring temperature reconstructions showed extremely little warming trends since the LIA and even a slight cooling during the second half of the 20th century.

The western Indian Himalayas were resisting global warming due to the emergence of a cooling trend in the late 20th century. There appears to have been a decreasing tendency in mean daily temperatures over the spring months as a result of the daily minimum and maximum temperatures. However, other tree-ring records from the same area demonstrated increased growth as a result of warmer winter temperatures since the mid-20th century. A large-scale composite of Himalayan spruce (*Picea spinulosa*) tree-ring data beginning in 1400 show a significant warming trend over the summer season there since 1700. The statistics show a clear below-average growth phase from 1590 to 1820, which is interpreted here as the LIA in Bhutan. This is followed by an almost monotonic growth trend up to the present. The correlation between this tree-ring record and the gridded summer temperatures indicates that this record can be understood as an indication of summer temperature change over Bhutan. The tree-ring data are centered exactly over the Himalayas of Bhutan, which is also the location with the largest concentration of significant correlations. The climatic records of Bhutan span only 10 to 20 years, making them inadequate for investigation. Therefore, it would seem that the current glacier retreat in Bhutan is related to the region of the Himalayas experiencing unusual summer warming. To show more important temporal changes, regional composites for the Tibetan Plateau have been created using decadal averages of oxygen isotopes during the previous 2,000 years. These Tibetan Plateau ice cores provide a 2,000-year perspective that reveals significant and unprecedented warming at high elevations(Chen et al., 2023). Whether the record is used as a substitution for temperature, precipitation, or atmospheric circulation, the oxygen isotopic record amply demonstrates the changes in large-scale dynamics over the past century. Ice core data gathered throughout the Tibetan Plateau, akin to tree ring chronologies, reveal the region to be a complex and climatically diverse area. Relatively unbiased records of the climate can be found in ice core data sets from exposed mountain tops far from the influences of topographic shielding and urbanization.

CHAPTER 10

PAST AND PRESENT GLACIAL ENVIRONMENTS

Contents

10.1 INTRODUCTION

Ice sheets and glaciers can be present in isolated areas that are far from populated areas. However, their inherent value and significance to each and every one of our lives should not be diminished by this physical distance. Our understanding of the current state of the global ecosystem as well as that of the past and future is critically dependent on glacial habitats. Glacial ecosystems appear complicated, and geologically transient at first glance, but underneath this is a pattern driven by processes. There is probably no other sedimentological environment that shows such quick, dynamic, and spatially variable changes in processes. These changes may appear fuzzy at small scales, but at larger scales, they start to show a degree of precision in the temporal and spatial organization and execution of individual processes. Global glaciations have occurred repeatedly throughout Earth's history as a result of the interplay of the oceans, continents, and atmosphere in response to solar radiation forcing. The geological stratigraphic record, which spans from the Precambrian to the Quaternary, has evidence of these glacial eras. The Ordovician in Mauritania, Saharan Africa, glacial striations and floating ice mass in the Sturtian of the Precambrian in South Australia, or fissure patterns in the diamictites of the Precambrian Gowganda in Ontario are examples of evidence that can be identified (Young & Nesbitt, 1985). These features are the result of the subglacial rearrangement of bedrock. Large-scale glacial sediments and related landforms can be found in mountainous regions, providing evidence of colder Pleistocene climate conditions in the past. There is ample evidence of the cold temperature intervals during the medieval era, also referred to as the Little Ice Age (LIA). For instance, the extremely severe winters and the following drought of that era were documented by painters, priests, and Icelandic writers throughout Europe. As in the past, glaciers and ice sheets have had a significant impact on all facets of earth systems, both positively and negatively. The intricacies of ice dynamics, together with the relevant factors and influences they have on the balance of ice mass and the particular causes of oscillations in the ice front and meltwater discharge, are still poorly understood.

10.2 INTERACTIONS OF GLACIAL FACTORS

The delicate interplay between ice motion and subglacial hydrology, as well as the temperature and sedimentological conditions of subglacial settings, remain little understood, as are the processes of ice basal movement. It is possible to think of current glaciers as living replicas of earlier ice sheets and glaciers. The modeling of contemporary ice masses and mass balance research contribute to our understanding of the historical growth and expansion of the global ice sheet. The various onshore and subaquatic environments of glaciers offer an active field laboratory for researching current glacial sedimentological processes, which can be used to comprehend and interpret glacial deposits from earlier geological periods. However, caution must be used when applying current ice masses as historical counterparts because they may have differed greatly in many important ways. It is increasingly evident that the Earth is basically a glacial planet interrupted by times of conditions equivalent to or occasionally warmer than the period we live in today, rather than viewing global glaciation as a series of environmental aberrations. The impact and repercussions of glaciation affect almost every area of life on Earth, either more or less. The direct effects of global glaciation can be seen in the distribution of flora, animals, early humans, soil types, and coastal topography, among other things. Climate and biogeographic changes have been brought about by the boundaries on land and the repeated variations in ocean sea level, even in the tropics, where conditions have remained nearly unchanged for the past 15 million years. These changes are a response to global glaciation. The impact of ice masses near humans has been extensively studied. Examples of these effects include meltwater surges and rapid ice advances that cause property loss, pasture land loss, and, in certain situations, human fatalities. Of course, these negative factors must be weighed against the advantages of having water for irrigation, hydropower projects, and fresh water supplies for homes. Less evident, but still being studied today, is the more delicate and widespread impact that modern ice masses have on ocean currents and global climate. To prepare for and anticipate sea level rise in the upcoming century, when estimates of global warming are expected to increase, we must expand our understanding of current glacial conditions.

10.3 GLACIATIONS

The importance of past glaciations and the sediments and forms they left behind cannot be overstated, given the scope of this influence. It is highly likely that there will be further global glaciations on Earth. Additionally, It is possible, though to what degree is yet unknown, that human activity has accelerated or worsened some of the intricate interactions between solar radiative forcing, and the oceanic/atmospheric circulation, particularly during the past 200 years. Future ice sheet stability is related to the possibility of sea level rise brought on by ice sheet melting in Greenland and Antarctica. Massive freshwater plumes could be pumped into the oceans if these ice sheets melt more quickly. This could have an impact on future ocean habitats, currents, surface ocean water temperatures such as those caused by El Nino–Southern Oscillation events, and ultimately, global weather patterns(Wang et al., 2017). There is still a debate regarding whether or not global warming is causing the faster melting of ice sheets. Creating comparisons to historical glacial conditions is another way that glacial environments aid in our understanding of the mechanisms involved in glacial sedimentation. A great deal of information on the origins, movements, and final deposition of glacial sediments on land and in water can be obtained by examining current glacial environments and sedimentological processes. At least 30% of the continental landmasses are currently covered by Pleistocene glacial sediments; this percentage must be increased when Pre-Pleistocene sediments, which span large portions of India, Australia, Africa, and South America, are taken into account. The location of landfills, the nutritional content of soils, the kind of groundwater supplies, the establishment of buildings, highways, and runways, as well as possible land routes for the disposal of contaminated waste, are all impacted by these sediments, which have an almost universal effect on human existence. These few instances highlight how important it is to comprehend the glacial processes that are taking place in glacial environments. This 30% estimate for Pleistocene sediments was accepted until quite recently; however, if glaciomarine sediments are taken into account, the current estimate may be closer to 60%. Large swaths of the northern and southern oceans of the Atlantic and Pacific are covered in these dense sediments that are spread out over the ocean floors. Although the influence of glaciomarine sediments on land-based habitats is undoubtedly small and only slightly affects fisheries, the significance of these vast regions of glacial sediments could grow if oceanic basins are used in the future.

There is evidence of multiple glacial periods in certain places with deep stratigraphic successions, which are separated by rocks that show signs of a warm environment and very strong weathering. A negative feedback mechanism is responsible for the cyclical fluctuations in temperature between cold and warm weather. The continental crust would have been uplifted and weathered more effectively as a result, lowering atmospheric CO_2. Lower temperatures would prevent weathering in areas covered with ice after continental glaciation began. Decreased rates of weathering would have finally resulted in a slow accumulation of CO_2 in the atmosphere, which would have destroyed the ice sheets and established a mild climate regime. Until it was disrupted by another event, like the breakup of the supercontinent, this long-term cycle of warm and cold climatic regimes would persist. If the early atmosphere's CO_2 content exceeded current levels, the commencement of global glaciation might have occurred at far higher partial pressures of CO_2. Following the Paleoproterozoic ice age, there was 1300 Ma during which there is no proof of glaciation (Hambrey & Harland, 1985). One or more supercontinents appear to have existed during the Mesoproterozoic era. Every continent experienced glaciers throughout the Neoproterozoic era. There may have been three peaks to this extensive and long-lasting glacial event, occurring at roughly 940, 770, and 615 Ma. In the region surrounding the North Atlantic, as well as in central and southern Africa, Brazil, western North America, and Australia, some of these are documented in the same stratigraphic sequence. Numerous glaciogenic successions from the Late Proterozoic show signs of concurrent rift activity. Certain locations with iron formations and sedimentary rocks rich in manganese have been explained as the result of hydrothermal activity connected to volcanism.

The equatorial belt would have seen preferred glaciation in the case of a drop in temperatures. The Late Proterozoic was a time of rapid plate tectonic movement, which may have allowed continents to move into high latitudes where they were glaciated. However, when continents moved into low paleolatitudes, their magnetic signature was not recorded on the sediments until much later. Adopting the idea of a supercontinent and its breakup in near-equatorial latitudes could have resulted in glacial through active chemical weathering and a decrease in atmospheric CO2. The polar seas would have frozen first during such a cooling phase, but gradually the continental region would have become glaciated. High-altitude regions would have been the starting point for glaciation, which would eventually have spread to sea level. As a result of the widespread glaciation and subsequent temperature drop, negative feedback would have caused atmospheric CO2 to once again build up. A sedimentological record showing evidence of alternating warm and cold climatic events would be predicted by such a system. Such glaciations would have occurred at low paleolatitudes. There is proof that glaciers existed during the Permo-Carboniferous, Devonian, and Ordovician periods. The majority of the evidence is located in the southern hemisphere continents that later joined to form the supercontinent Gondwana. The movement of distinct portions of the supercontinent over the south polar region and the several glaciations of the Permo-Carboniferous seem to be related. The glaciation of the Late Paleozoic era may have been caused by diverse major processes. Given that a portion of the supercontinent was at high latitudes, the emergence of the Gondwana supercontinent may have contributed to atmospheric CO2 depletion, which would have lowered average summer temperatures, caused snow to accumulate, and facilitated the formation of continental glaciers.

10.4 PALEOGEOGRAPHY AND GLACIERS

Like the Precambrian, the early Paleozoic paleogeography differed from the current conditions (Swanson-Hysell, 2021). The majority of the continents, North America, Europe, and Siberia were found in tropical or subtropical regions. But during the Paleozoic, the Gondwanaland supercontinent stretched between the equator and the South Pole, engulfing parts of South America, Africa, Arabia, Madagascar, India, Australia, and Antarctica in addition to Florida, southern and central Europe, Turkey, Iran, Afghanistan, Tibet, and New Zealand. The South Pole drift was generally followed by the Paleozoic glacial centers. According to continental reconstructions for the Late Ordovician, the Gondwana continents were united while the northern continents were divided. North Africa was near the South Pole, which is where the best remnants of the Late Ordovician glaciers are found. Glaciation in the Devonian and Permo-Carboniferous periods appears to have developed similarly to that of the Ordovician–Silurian periods, covering a period of nearly 100 Ma from roughly 360 to 255 Ma when Gondwanaland started to drift away from the South Pole. There have been reports of Devonian rocks from the Amazon basin, Brazil, West Africa, and Antarctica that may have originated from glaciers. In Africa and Australia, evidence of this extensive pre-Late Cenozoic ice period was first found in 1859. Paleolatitudes typically lie between 48 and 80°S. Nonetheless, there are also documented Permian glaciogenic units from northeastern Siberia, which at the time was close to the North Pole. Large-scale ice sheet glaciation spanned most of Gondwanaland, including India, starting at 330 Ma in South America and southwest Africa. However, this glaciation was most prominent in Africa, where the Dwyka Formation, which dates to approximately 280 Ma, describes a subpolar, marine-based ice sheet. The early Permian period is when the glaciogenic Talchir Boulder Beds in India were deposited. Hot deserts formed in eastern South America and northern Africa, which at the time were in tropical latitudes, concurrently with the end of the glaciers in Australia. During this period, the Gondwanaland and Laurasian supercontinents fused to form Pangaea, resulting in massive lava flows and the formation of mountain ranges.

In comparison to the Phanerozoic, the Precambrian glaciation appears to have been more widespread and persistent. It seems improbable that a single factor causes all glaciations; instead, a complex interaction of factors at the Earth's surface and within the planet is most likely responsible. This is explained by high CO2 levels and a minute amount of exposed continental crust. The first extensive glaciations in the Paleoproterozoic have been linked to a decrease in CO2 and an increase in continental crust. While there is evidence for the existence of a supercontinent, glaciation does not occur throughout the extended period between approximately 2.0 and 1.0 Ga (Calkin & Young, 2002). This is linked to an extended period of intense volcanic activity, which is thought to have kept atmospheric CO2 levels high and prevented glaciation. The world's largest known ice sheet proliferation can be seen in the Neoproterozoic record. These glacial centers evolved at low paleolatitudes in several cases. The idea that a supercontinent or supercontinents existed during this time is supported by some paleomagnetic data and extensive evidence of drifting. Large-scale freezing into low paleolatitudes and an unparalleled drop in atmospheric CO2 could have resulted from such a supercontinental structure at tropical latitudes. A return to a warm climatic regime and a warming trend would have been finally generated by a negative feedback mechanism. This would have allowed for the production of a warming and cooling climate. When continents split at or close to the end of the Precambrian, the cycle of warm and cold climates was disrupted; nonetheless, no clear cycle can be identified for these ancient ice ages. Paleontological data from the Phanerozoic are considerably large. Continents in high paleolatitudes appear to have experienced these recent glaciations. High-latitude regions were impacted by the Cenozoic glaciation and potentially the Late Ordovician event at times of low sea level. There are no tillites identified from Jurassic and Cretaceous units, nor are there known glaciogenic deposits from the Triassic, while the possibility of substantial glacier cover across Siberia and Antarctica at high paleolatitudes cannot be ruled out. Due to the continuous breakup of the supercontinents Gondwanaland and Laurasia, which resulted in high sea levels, widespread continental floods, and the formation of new ocean basins, the Mesozoic era saw equal warming of the planet. Temperatures 10–15°C above current levels at high latitudes and CO2 concentrations five times higher than those of the pre-industrial era were the main characteristics of this and likely previous non-glacial climatic regimes.

10.5 PAST GLACIATIONS

Paleogeographic shifts of significant magnitude were occurring during the Mesozoic and early Tertiary periods. By the Tertiary Period, Antarctica had migrated into the region around the South Pole, while Australia had begun to travel northward from Antarctica. The deepening of the Drake Passage between Antarctica and South America during the Late Oligocene to Miocene period marked the beginning of a significant phase in Antarctica's thermal isolation. The North Atlantic was opening up in the northwest and northeast at the same time that the Arctic Basin was developing due to the North Atlantic Ridge's spread. This made it possible for Greenland to split off from Europe at about 37 Ma, which resulted in the creation of deep-water flow in the North Atlantic. The Middle to Late Eocene must have seen intense cooling in Antarctica, resulting in the accumulation of mountain ice and winter snow cover between 52 and 36 Ma. An extensive, full-scale ice sheet had formed by at least the early Oligocene (–36 Ma). Interpretations of the Antarctic glacial record that imply significant fluctuations of the ice sheets may have occurred at intervals of as little as one to two million years over the last 36 Ma, and that the ice cover was likely more extensive than it is today. Following the opening of the Drake Passage, there was a notable expansion of the East Antarctic Ice Sheet throughout the Middle Miocene. West Antarctic ice shelves began to form as a result of ongoing cooling, and by 10 Ma, the West Antarctic Ice Sheet was finally placed below sea level(Vaughan, 2008). Further deglaciation may have been triggered by a warm Early Pliocene, but the Transantarctic Mountains saw a period of fast uplift that may have coincided with the beginning of Late Pliocene glacier readvances in the northern hemisphere around 2.5–2.4 Ma. Mountain glaciers or ice caps in the mid-latitude Patagonian Andes were more extensive than they are today elsewhere in the southern hemisphere due to Miocene uplift and cooling by 4.7 Ma. Although there is no direct evidence of the Greenland Ice Sheet growing in the northern hemisphere, mountain glaciers in the mountains of East and West Greenland were formed by 10–2.7 Ma. Over the continent, an ice sheet at least as large as it is today was most likely formed by 2.4 Ma. In Iceland, small-scale glaciation dates back to around 3 Ma. Understanding the nature, extent, and chronology of the Late Cenozoic glaciation has been made possible by the creation of new, comprehensive records in boths glaciated and unglaciated regions as well as significant adjustments to the traditional glacial–interglacial chronologies. For instance, radiocarbon dating made it possible to estimate changes in continental ice volume and climate while concentrating on specifics of the most recent Quaternary glacial oscillations. Similar global glacial and climatic inferences were drawn at the same period by utilizing deep cores, which were developed in 1947. These made it possible to gather undisturbed records of ocean bottom sediments that were accumulated over extended periods at rather steady rates. The stratigraphy and subsequent chronologies derived from deep-sea cores, which comprise the acquisition of the oxygen-isotope ratios of microorganisms, namely the calcareous planktonic and benthic foraminifera, may be the primary source of our understanding of ice ages. For the Late Cenozoic, cores offered a continuous time series of the global fluctuations in continental ice volume that are typically not reproducible by the shorter and more fragmented continental records. Ocean mixing times are relatively short, which accounts for the patterns in isotopic ratio fluctuations found in deep sea cores that are almost synchronous globally and easily correlated based on distinct stratigraphic strata. As a result, they have been used as a benchmark for measuring other marine and continental chronologies. These deep-sea isotopic records show that up to 16 or more major ice advances and retreats occurred during the Pleistocene and an even greater number when the record is extended into the Pliocene, instead of the four old major glaciations that differentiated in central North America or Europe through the mid to late twentieth century. Deep sea cores can also be used to estimate the size of ice sheets on adjacent continents by measuring the amount of continental debris that is incorporated, estimating the sea surface temperature (SST) using biological transformations, and calculating the percentage of $CaCO_3$ that is controlled by the depth (Shen et al., 1996).

10.6 LOESS

Loess deposition is primarily a Quaternary phenomenon. Very large loess deposits in central Asia provide more comprehensive terrestrial paleoclimatic records for the Late Cenozoic. Furthermore, many lacustrine sequences reflect the Pleistocene along with the sediments and pollen records. Many of these records can also be directly correlated with the glacial–interglacial 18O changes of deep-sea sediment records, which provide details about the Middle Pleistocene. These records are dated using techniques such as faunal changes, carbon–14, thermoluminescence, and paleomagnetic measurements. Loess is carried and deposited by wind in unconsolidated and unstratified accumulations. A sequence of alternating soil strata and loess totaling over 200 meters displays a record of global climate oscillations similar to that found in deep sea cores at sites in central China, including Xifeng, and Xian. Throughout the core, this sequence demonstrates recurrent, climatic oscillations that last for several eras. The basins of the tropical Andes, southern Japan, the Dead Sea, and northern California are the sources of long lacustrine cores. The oxygen isotope stages of deep-sea cores that are assigned to specific glaciations serve as the main source. Tills or proglacial deposits were the main markers for glacial advance and retreat in the northern hemisphere. Because river terraces are linked to glacier deposits, river terraces served as the foundation for glacial advances in the European Alps.

10.7 ICE SHEET

There is no evidence of a Scandinavian Ice Sheet advance during the Early Pleistocene, with the possible exception of some advances that have been documented in west-central and the neighboring Commonwealth of Independent States. There are reports of early Middle Pleistocene glaciations from Russia, Belarus, Poland, Ukraine, and Britain. There is a strong correlation between the Middle and Late Pleistocene glaciations and the corresponding events in North America. The firmness of the Antarctic Ice Sheet in the Cenozoic will be significantly impacted by how this formation is interpreted. The huge Mid-Miocene ice sheet later shrank to its current extent. One of the best-preserved and most well-dated terrestrial glacial records in the world may be found in the Southern (Patagonian) Andes, which experienced many glacial ice sheets during the Late Cenozoic. Glaciers that originated as early as the Late Miocene and before 4.7 Ma left behind many drifts on the mountain slopes during the Pliocene, which may be correlated with deposits from the Peruvian–Bolivian Andes and New Zealand that are not as well-dated. During the Early Pleistocene, when there was less glacial in the northern hemisphere, an extensive glacier in the Patagonian region took place. Southern Patagonian glaciers reached the Atlantic Continental Shelf 200 km east of the highlands at this epoch, 1.2–1.0 Ma (Steinthorsdottir et al., 2021). It is assumed that throughout the Late Middle Pleistocene, glacial advances took place over the southern hemisphere mountain range. The records of the Greenland and Antarctic ice cores demonstrate the well-documented phase of the glacial cycles of the ice sheets in the northern and southern hemispheres. Synchronization of global glacial climatic oscillations with those in Greenland was revealed by δ18O ice core data from Taylor Dome, close to coastal Antarctica. The maximum of the southern edges of the northern hemisphere ice sheets coincides with the ice sheet advance across the Ross Sea, based on the data from the Transantarctic Mountains and the Ross Sea in Antarctica. The expansion and disintegration of the terrestrial, northern hemisphere ice sheets directly affects the variations of the Antarctic Ice Sheet. This direct effect is related to the interaction between the growth and decay of continental ice sheets and variations in the global sea level. Sea level recession (Northern hemisphere) causes the Antarctic Ice Sheet to expand because a large portion of it is below sea level; on the other hand, Antarctic ice retreat results in increased calving and disintegration concerning sea level rise.

10.8 LITTLE ICE AGE

The Little Ice Age is generally accepted to have culminated in the subsequent decline in temperatures caused by decreased summer insolation. On top of these extensive Holocene climate shifts, there have been shorter-term temperature fluctuations accompanied by advances and retreats of mountain glaciers. The Little Ice Age, a global phenomenon characterized by temperature fluctuations, seems to have started in the thirteenth and fourteenth centuries with small glacial advances. After a brief period of mild weather, it entered a phase of recurring glacial expansions and retreats around 1600 AD, which culminated in the last 200 years in global warming and retreat of temperate and subpolar mountain glaciers.

The availability of accurate tree ring dating and the many historical observations in Europe, in particular, implies that glacial fluctuations occurred almost simultaneously over several regions extending over several hundred kilometers. Heat distribution over the Earth's surface and ocean circulation can both be significantly impacted by plate tectonic placement of the continents. Changes in the oceanic circulation can also be used to explain variations in the climate. The solar flux has increased by around 1% since the Permo-Carboniferous Ice Age (Macarewich & Poulsen, 2022).

Additionally, the continents have connected and divided again, moving to their current positions, and plateau and mountain altitudes have increased. Significant impacts on ocean temperatures and salinity, rainfall, temperature patterns, and atmospheric composition must have resulted from modifications in oceans, atmospheric vapor transport, and wave routes. Research indicates that CO_2 concentrations were five to ten times higher in the Mesozoic than they are now. The overall, stepwise cooling during this period can be explained by the apparent decline in atmospheric CO_2 from the Late Cretaceous through the Pleistocene and the resulting decreased greenhouse warming. However, it is unclear what is causing the rapid cooling and fluctuations in CO_2. Volcanic outgassing and declining global sea-floor spreading rates have been attributed to the drop in atmospheric CO_2 levels. A primary factor in climate change may have also been decreased global chemical weathering, which lowers CO_2, which raises sea levels and reduces land area.

It has long been believed that the shift of land masses toward polar latitudes is essential to the onset of continental glaciation. Nonetheless, paleogeographic reconstructions show that throughout the previous 100 million years, Antarctica and the northern continents have been close to their current latitudes. In the past, it was thought that local geographic variables had a crucial role in the development of Late Cenozoic glaciers at high latitudes. These included the development of the Isthmus of Panama, the lifting of ice-sheet centers in Arctic Canada, the opening of the Drake Passage, which isolated Antarctica from South America, and the potential rerouting of important ocean currents to northerly routes. Furthermore, it has not been demonstrated that enhanced Cenozoic volcanism and solar shielding have supplied sufficient climatic forcing for Antarctic and mid-latitude northern hemisphere glaciations at the appropriate times.

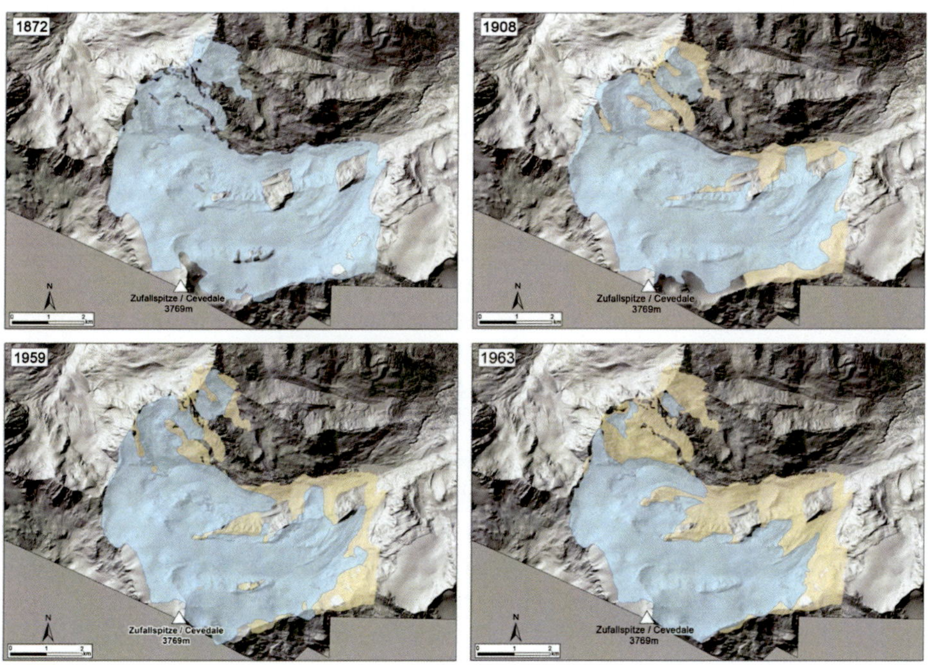

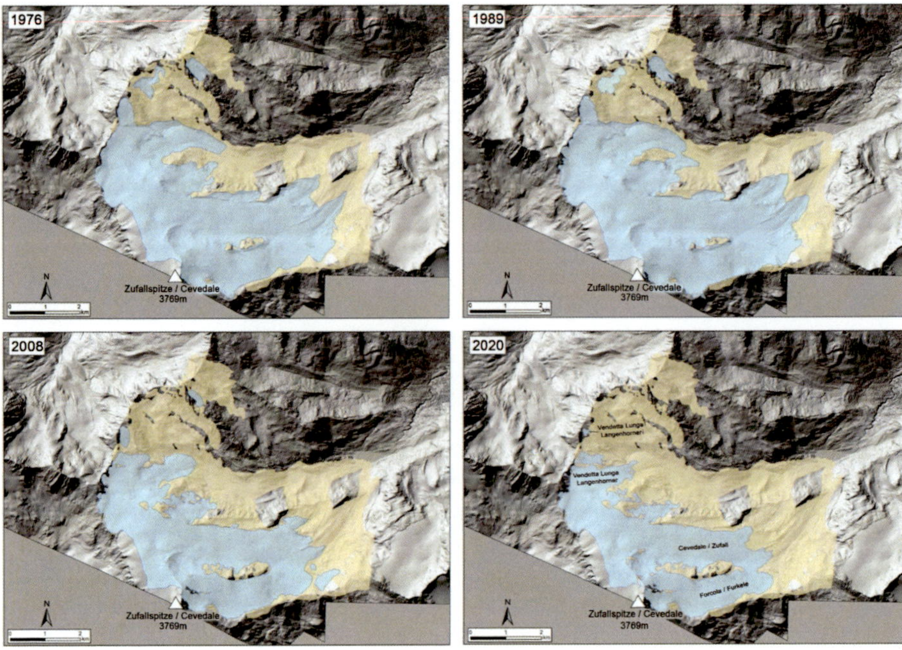

Figure 10.1 Evolution of the glacierized area (light blue) and proglacial area (yellow) at the head of Martello Valley. Glacier limits reconstructed after manual digitalization of available topographic maps and orthophotos. Toponyms in the last box refer to the New Italian Glacier Inventory

Source: Azzoni RS, Pelfini M, Zerboni A. Estimating the Evolution of a Post-Little Ice Age Deglaciated Alpine Valley through the DEM of Difference (DoD). Remote Sensing. 2023; 15(12):3190. https://doi.org/10.3390/rs15123190

According to models of uplift in the northern hemisphere, the upper atmospheric jet stream route would have changed the Rossby wave pattern, facilitating the penetration of cold arctic air mass over more southern regions of the northern hemisphere continents. During the Late Neogene, the elevation of the Tibetan Plateau, the Himalayas, and most likely the Andes caused a sharp rise in weathering, which in turn drastically reduced atmospheric CO2 and caused global cooling. The Holocene or the Pleistocene–Holocene transition are examples of short-term climatic shifts that cannot be explained by longer-term temporal cycles of Milankovitch-type orbital perturbations. It appears likely that CO2 mechanisms or the ocean-atmosphere had an impact, at least in the early Holocene. The need to forecast future climatic change and the expansion of ice sheets grows with our understanding of past global climates and the mechanisms underlying climatic change. Such forecasts need to be framed in terms of the influence of human civilizations on historical and contemporary climates. Based on analysis of glaciation and climate change during the Late Cenozoic, it seems that the Earth is currently going through an interglacial period (Hay et al., 2002). There are no clear signs of an imminent glacier based on short-term climate patterns. There is historical and geological data that shows the growth and expansion of glaciers in the Arctic during times of global warming. Increased calving along the ice margin with sea level rise is a response to increased precipitation during the Holocene in Antarctica. Even the most advanced computer models won't be able to predict the climate until more paleoclimatic data as well as knowledge on the chemical, physical, and biological interactions of the modern atmosphere and oceans are collected. Medium to high stresses are of greater importance in glacial geology, and not all issues can always be resolved using a mechanical approach to the soil. The current ignorance of time-dependent deformation mechanisms in unlithified sediments is the most pressing issue. A thorough understanding of soft sediment deformation mechanisms is essential, as reconstructing past ice sheet conditions is one of the key objectives of glaciotectonic investigations. When shear pressures are greater than its shear strength, a sediment deforms.

This book has given an introduction to the various aspects of glaciers and their associated processes that can be used by students and researchers in the field.

REFERENCES

1. Aber, J. S., Croot, D. G., & Fenton, M. M. (1989). Large Composite-Ridges. In J. S. Aber, D. G. Croot, & M. M. Fenton (Eds.), Glaciotectonic Landforms and Structures (pp. 29–46). Springer Netherlands. https://doi.org/10.1007/978-94-015-6841-8_3
2. Alley, R. B., & Bindschadler, R. A. (2001). The West Antarctic Ice Sheet and Sea-Level Change. In The West Antarctic Ice Sheet: Behavior and Environment (pp. 1–11). American Geophysical Union (AGU). https://doi.org/10.1029/AR077p0001
3. Allison, I., Paul, F., Colgan, W., & King, M. (2021). Chapter 20—Ice sheets, glaciers, and sea level. In W. Haeberli & C. Whiteman (Eds.), Snow and Ice-Related Hazards, Risks, and Disasters (Second Edition) (pp. 707–740). Elsevier. https://doi.org/10.1016/B978-0-12-817129-5.00013-5
4. Andersen, K. K., Svensson, A., Johnsen, S. J., Rasmussen, S. O., Bigler, M., Röthlisberger, R., Ruth, U., Siggaard-Andersen, M.-L., Peder Steffensen, J., Dahl-Jensen, D., Vinther, B. M., & Clausen, H. B. (2006). The Greenland Ice Core Chronology 2005, 15–42 ka. Part 1: Constructing the time scale. Quaternary Science Reviews, 25(23), 3246–3257. https://doi.org/10.1016/j.quascirev.2006.08.002
5. Anderson, R. S., Anderson, L. S., Armstrong, W. H., Rossi, M. W., & Crump, S. E. (2018). Glaciation of alpine valleys: The glacier – debris-covered glacier – rock glacier continuum. Geomorphology, 311, 127–142. https://doi.org/10.1016/j.geomorph.2018.03.015
6. Arnold, N., Richards, K., Willis, I., & Sharp, M. (1998). Initial results from a distributed, physically based model of glacier hydrology. Hydrological Processes, 12(2), 191–219. https://doi.org/10.1002/(SICI)1099-1085(199802)12:2<191::AID-HYP571>3.0.CO;2-C

7. Axford, Y., Geirsdóttir, Á., Miller, G. H., & Langdon, P. G. (2009). Climate of the Little Ice Age and the past 2000 years in northeast Iceland inferred from chironomids and other lake sediment proxies. Journal of Paleolimnology, 41(1), 7–24. https://doi.org/10.1007/s10933-008-9251-1
8. Azam, Mohd. F., Kargel, J. S., Shea, J. M., Nepal, S., Haritashya, U. K., Srivastava, S., Maussion, F., Qazi, N., Chevallier, P., Dimri, A. P., Kulkarni, A. V., Cogley, J. G., & Bahuguna, I. (2021). Glaciohydrology of the Himalaya-Karakoram. Science, 373(6557), eabf3668. https://doi.org/10.1126/science.abf3668
9. Aziz, A. (2000). Chapter 16—Stratigraphy and hydrocarbon potential of the Lower Palaeozoic succession of License NC-115, Murzuq Basin, SW Libya. In M. A. Sola & D. Worsley (Eds.), Geological Exploration in Murzuq Basin (pp. 349–368). Elsevier Science B.V. https://doi.org/10.1016/B978-044450611-5/50018-0
10. Belkin, I. M., Levitus, S., Antonov, J., & Malmberg, S.-A. (1998). "Great Salinity Anomalies" in the North Atlantic. Progress in Oceanography, 41(1), 1–68. https://doi.org/10.1016/S0079-6611(98)00015-9
11. Bellwald, B., Stokke, H., Winsborrow, M., Planke, S., Sættem, J., Lebedeva-Ivanova, N., Hafeez, A., Kurjanski, B., Myklebust, R., & Polteau, S. (2023). Structural and fluid-migration control on hill-hole pair formation: Evidence from high-resolution 3D seismic data from the SW Barents Sea. Geomorphology, 420, 108502. https://doi.org/10.1016/j.geomorph.2022.108502
12. Benediktsson, Í. Ö., Möller, P., Ingólfsson, Ó., van der Meer, J. J. M., Kjær, K. H., & Krüger, J. (2008). Instantaneous end moraine and sediment wedge formation during the 1890 glacier surge of Brúarjökull, Iceland. Quaternary Science Reviews, 27(3), 209–234. https://doi.org/10.1016/j.quascirev.2007.10.007
13. Benn, D. I., Bolch, T., Hands, K., Gulley, J., Luckman, A., Nicholson, L. I., Quincey, D., Thompson, S., Toumi, R., & Wiseman, S. (2012). Response of debris-covered glaciers in the Mount Everest region to recent warming, and implications for outburst flood hazards. Earth-Science Reviews, 114(1), 156–174. https://doi.org/10.1016/j.earscirev.2012.03.008
14. Benn, D. I., Owen, L. A., Osmaston, H. A., Seltzer, G. O., Porter, S. C., & Mark, B. (2005). Reconstruction of equilibrium-line altitudes for tropical and sub-tropical glaciers. Quaternary International, 138–139, 8–21. https://doi.org/10.1016/j.quaint.2005.02.003
15. Bennett, M. R. (2001). The morphology, structural evolution and significance of push moraines. Earth-Science Reviews, 53(3), 197–236. https://doi.org/10.1016/S0012-8252(00)00039-8

16. Bennett, M. R. (2003). Ice streams as the arteries of an ice sheet: Their mechanics, stability and significance. Earth-Science Reviews, 61(3), 309–339. https://doi.org/10.1016/S0012-8252(02)00130-7
17. Bennett, M. R., Huddart, D., & Glasser, N. F. (1999). Large-scale Bedrock Displacement by Cirque Glaciers. Arctic, Antarctic, and Alpine Research, 31(1), 99–107. https://doi.org/10.1080/15230430.1999.12003285
18. Bhattacharya, B. K., Mallick, K., Patel, N. K., & Parihar, J. S. (2010). Regional clear sky evapotranspiration over agricultural land using remote sensing data from Indian geostationary meteorological satellite. Journal of Hydrology, 387(1), 65–80. https://doi.org/10.1016/j.jhydrol.2010.03.030
19. Bickerdike, H. L., Evans, D. J. A., Stokes, C. R., & Ó Cofaigh, C. (2018). The glacial geomorphology of the Loch Lomond (Younger Dryas) Stadial in Britain: A review. Journal of Quaternary Science, 33(1), 1–54. https://doi.org/10.1002/jqs.3010
20. Björnsson, H. (2003). Subglacial lakes and jökulhlaups in Iceland. Global and Planetary Change, 35(3), 255–271. https://doi.org/10.1016/S0921-8181(02)00130-3
21. Blatter, H., & Greve, R. (2015). Comparison and verification of enthalpy schemes for polythermal glaciers and ice sheets with a one-dimensional model. Polar Science, 9(2), 196–207. https://doi.org/10.1016/j.polar.2015.04.001
22. Bol'shakov, V. A. (2014). A link between global climate variability in the Pleistocene and variations in the Earth's orbital parameters. Stratigraphy and Geological Correlation, 22(5), 538–551. https://doi.org/10.1134/S0869593814050049
23. Borgne, P. L., Legendre, G., & Marsouin, A. (2004). Meteosat and GOES-East Imager Visible Channel Calibration. Journal of Atmospheric and Oceanic Technology, 21(11), 1701–1709. https://doi.org/10.1175/JTECHJTECH-1675.1
24. Boulton, G. S. (1982). Processes and Patterns of Glacial Erosion. In D. R. Coates (Ed.), Glacial Geomorphology (pp. 41–87). Springer Netherlands. https://doi.org/10.1007/978-94-011-6491-7_2
25. Braathen, A., Osmundsen, P. T., Hauso, H., Semshaug, S., Fredman, N., & Buckley, S. J. (2013). Fault-induced deformation in a poorly consolidated, siliciclastic growth basin: A study from the Devonian in Norway. Tectonophysics, 586, 112–129. https://doi.org/10.1016/j.tecto.2012.11.008
26. Brown, C. W., Connor, L. N., Lillibridge, J. L., Nalli, N. R., & Legeckis, R. V. (2005). An Introduction to Satellite Sensors, Observations and

Techniques. In R. L. Miller, C. E. Del Castillo, & B. A. Mckee (Eds.), Remote Sensing of Coastal Aquatic Environments: Technologies, Techniques and Applications (pp. 21–50). Springer Netherlands. https://doi.org/10.1007/978-1-4020-3100-7_2

27. Brown, G. H. (2002). Glacier meltwater hydrochemistry. Applied Geochemistry, 17(7), 855–883. https://doi.org/10.1016/S0883-2927(01)00123-8
28. Brunet, T. C., García, J. M. V., & González, V. M. H. & A. U. (2008). GIS landscape models for the study of preindustrial settlement patterns in Mediterranean areas. In Geoinformation Technologies for Geo-Cultural Landscapes: European Perspectives. CRC Press.
29. Calkin, P. E., & Young, G. M. (2002). 2—Global glacial chronologies and causes of glaciation. In J. Menzies (Ed.), Modern and Past Glacial Environments (pp. 15–52). Butterworth-Heinemann. https://doi.org/10.1016/B978-075064226-2/50005-2
30. Cape, M. R., Vernet, M., Kahru, M., & Spreen, G. (2014). Polynya dynamics drive primary production in the Larsen A and B embayments following ice shelf collapse. Journal of Geophysical Research: Oceans, 119(1), 572–594. https://doi.org/10.1002/2013JC009441
31. Carling, P. A. (2013). Freshwater megaflood sedimentation: What can we learn about generic processes? Earth-Science Reviews, 125, 87–113. https://doi.org/10.1016/j.earscirev.2013.06.002
32. Carozzi, A. V. (1984). Glaciology and the Ice Age. Journal of Geological Education. https://www.tandfonline.com/doi/abs/10.5408/0022-1368-32.3.158
33. Castelnau, O., Duval, P., Montagnat, M., & Brenner, R. (2008). Elastoviscoplastic micromechanical modeling of the transient creep of ice. Journal of Geophysical Research: Solid Earth, 113(B11). https://doi.org/10.1029/2008JB005751
34. Castelnau, O., Shoji, H., Mangeney, A., Milsch, H., Duval, P., Miyamoto, A., Kawada, K., & Watanabe, O. (1998). Anisotropic behavior of GRIP ices and flow in Central Greenland. Earth and Planetary Science Letters, 154(1), 307–322. https://doi.org/10.1016/S0012-821X(97)00193-3
35. Cawse-Nicholson, K., Townsend, P. A., Schimel, D., Assiri, A. M., Blake, P. L., Buongiorno, M. F., Campbell, P., Carmon, N., Casey, K. A., Correa-Pabón, R. E., Dahlin, K. M., Dashti, H., Dennison, P. E., Dierssen, H., Erickson, A., Fisher, J. B., Frouin, R., Gatebe, C. K., Gholizadeh, H., … Zhang, Q. (2021). NASA's surface biology and geology designated observable: A perspective on surface imaging algorithms. Remote Sensing of Environment, 257, 112349. https://doi.org/10.1016/j.rse.2021.112349

36. Chatterjee, R. S., Singh, N., Thapa, S., Sharma, D., & Kumar, D. (2017). Retrieval of land surface temperature (LST) from landsat TM6 and TIRS data by single channel radiative transfer algorithm using satellite and ground-based inputs. International Journal of Applied Earth Observation and Geoinformation, 58, 264–277. https://doi.org/10.1016/j.jag.2017.02.017
37. Chen, B., Xu, X.-D., Yang, S., & Zhang, W. (2012). On the origin and destination of atmospheric moisture and air mass over the Tibetan Plateau. Theoretical and Applied Climatology, 110(3), 423–435. https://doi.org/10.1007/s00704-012-0641-y
38. Chen, Z., Liu, J., Rühland, K. M., Zhang, J., Zhang, K., Kang, W., Chen, S., Wang, R., Zhang, H., & Smol, J. P. (2023). Collapse of the Tibetan Empire attributed to climatic shifts: Paleolimnological evidence from the western Tibetan Plateau. Quaternary Science Reviews, 317, 108280. https://doi.org/10.1016/j.quascirev.2023.108280
39. Chumakov, N. M. (2009). Chapter 11.2 Neoproterozoic Glacial Events in Eurasia. In C. Gaucher, A. N. Sial, H. E. Frimmel, & G. P. Halverson (Eds.), Developments in Precambrian Geology (Vol. 16, pp. 389–403). Elsevier. https://doi.org/10.1016/S0166-2635(09)01627-2
40. Cook, S. J., & Swift, D. A. (2012). Subglacial basins: Their origin and importance in glacial systems and landscapes. Earth-Science Reviews, 115(4), 332–372. https://doi.org/10.1016/j.earscirev.2012.09.009
41. Cracknell, A. P. (2018). The development of remote sensing in the last 40 years. International Journal of Remote Sensing, 39(23), 8387–8427. https://doi.org/10.1080/01431161.2018.1550919
42. Crucifix, M. (2018). Pleistocene Glaciations. In Climate Changes in the Holocene: CRC Press.
43. Dansgaard, W., Clausen, H. B., Gundestrup, N., Johnsen, S. J., & Rygner, C. (1985). Dating and Climatic Interpretation of Two Deep Greenland Ice Cores. In Greenland Ice Core: Geophysics, Geochemistry, and the Environment (pp. 71–76). American Geophysical Union (AGU). https://doi.org/10.1029/GM033p0071
44. Davies, B. J., Hambrey, M. J., Smellie, J. L., Carrivick, J. L., & Glasser, N. F. (2012). Antarctic Peninsula Ice Sheet evolution during the Cenozoic Era. Quaternary Science Reviews, 31, 30–66. https://doi.org/10.1016/j.quascirev.2011.10.012
45. DEMOREST, M. (1943). Ice Sheets. GSA Bulletin, 54(3), 363–400. https://doi.org/10.1130/GSAB-54-363

46. Denton, G. H. (2000). Does an asymmetric thermohaline–ice-sheet oscillator drive 100 000-yr glacial cycles? Journal of Quaternary Science, 15(4), 301–318. https://doi.org/10.1002/1099-1417(200005)15:4<301::AID-JQS536>3.0.CO;2-Z
47. Derbyshire, E., Shi Yafeng, Li Jijun, Zheng Benxing, Li Shijie, & Wang Jingtai. (1991). Quaternary glaciation of tibet: The geological evidence. Quaternary Science Reviews, 10(6), 485–510. https://doi.org/10.1016/0277-3791(91)90042-S
48. Dowdeswell, J. A., Hagen, J. O., Björnsson, H., Glazovsky, A. F., Harrison, W. D., Holmlund, P., Jania, J., Koerner, R. M., Lefauconnier, B., Ommanney, C. S. L., & Thomas, R. H. (1997). The Mass Balance of Circum-Arctic Glaciers and Recent Climate Change. Quaternary Research, 48(1), 1–14. https://doi.org/10.1006/qres.1997.1900
49. Dowdeswell, J. A., & Jeffries, M. O. (2017). Arctic Ice Shelves: An Introduction. In L. Copland & D. Mueller (Eds.), Arctic Ice Shelves and Ice Islands (pp. 3–21). Springer Netherlands. https://doi.org/10.1007/978-94-024-1101-0_1
50. Dwivedi, R. S. (Ed.). (2018). Active Remote Sensing. In Geospatial Technologies for Land Degradation Assessment and Management. CRC Press.
51. Ehlers, J., & Gibbard, P. L. (2007). The extent and chronology of Cenozoic Global Glaciation. Quaternary International, 164–165, 6–20. https://doi.org/10.1016/j.quaint.2006.10.008
52. Elzinga, A. (2017). Polar ice cores: Climate change messengers. In Research Objects in their Technological Setting. Routledge.
53. Evans, D. J. A. (2009). Controlled moraines: Origins, characteristics and palaeoglaciological implications. Quaternary Science Reviews, 28(3), 183–208. https://doi.org/10.1016/j.quascirev.2008.10.024
54. Evans, D. J. A., Roberts, D. H., & Evans, S. C. (2016). Multiple subglacial till deposition: A modern exemplar for Quaternary palaeoglaciology. Quaternary Science Reviews, 145, 183–203. https://doi.org/10.1016/j.quascirev.2016.05.029
55. Evans, I. S. (1977). World-Wide Variations in the Direction and Concentration of Cirque and Glacier Aspects. Geografiska Annaler: Series A, Physical Geography, 59(3–4), 151–175. https://doi.org/10.1080/04353676.1977.11879949
56. Faria, S. H. (2006). Creep and recrystallization of large polycrystalline masses. III. Continuum theory of ice sheets. Proceedings of the Royal Society A: Mathematical, Physical and Engineering Sciences, 462(2073),

2797–2816. https://doi.org/10.1098/rspa.2006.1698

57. Forsberg, R., Sørensen, L., & Simonsen, S. (2017). Greenland and Antarctica Ice Sheet Mass Changes and Effects on Global Sea Level. In A. Cazenave, N. Champollion, F. Paul, & J. Benveniste (Eds.), Integrative Study of the Mean Sea Level and Its Components (pp. 91–106). Springer International Publishing. https://doi.org/10.1007/978-3-319-56490-6_5
58. Fraser, C. S., Ogleby, C. L., & Collier, P. A. (1995). Evaluation of a Small Format Aerial Mapping System. Australian Surveyor, 40(1), 10–15. https://doi.org/10.1080/00050331.1995.10558488
59. Frezzotti, M., & Orombelli, G. (2014). Glaciers and ice sheets: Current status and trends. Rendiconti Lincei, 25(1), 59–70. https://doi.org/10.1007/s12210-013-0255-z
60. Galton-Fenzi, B. K., Hunter, J. R., Coleman, R., Marsland, S. J., & Warner, R. C. (2012). Modeling the basal melting and marine ice accretion of the Amery Ice Shelf. Journal of Geophysical Research: Oceans, 117(C9). https://doi.org/10.1029/2012JC008214
61. Goddéris, Y., Donnadieu, Y., Le Hir, G., Lefebvre, V., & Nardin, E. (2014). The role of palaeogeography in the Phanerozoic history of atmospheric CO2 and climate. Earth-Science Reviews, 128, 122–138. https://doi.org/10.1016/j.earscirev.2013.11.004
62. Gomez, B., Dowdeswell, J. A., & Sharp, M. (1988). Microstructural control of quartz sand grain shape and texture: Implications for the discrimination of debris transport pathways through glaciers. Sedimentary Geology, 57(1), 119–129. https://doi.org/10.1016/0037-0738(88)90021-8
63. Gordeev, V. V. (2006). Fluvial sediment flux to the Arctic Ocean. Geomorphology, 80(1), 94–104. https://doi.org/10.1016/j.geomorph.2005.09.008
64. Hald, M., Andersson, C., Ebbesen, H., Jansen, E., Klitgaard-Kristensen, D., Risebrobakken, B., Salomonsen, G. R., Sarnthein, M., Sejrup, H. P., & Telford, R. J. (2007). Variations in temperature and extent of Atlantic Water in the northern North Atlantic during the Holocene. Quaternary Science Reviews, 26(25), 3423–3440. https://doi.org/10.1016/j.quascirev.2007.10.005
65. Hallet, B., Hunter, L., & Bogen, J. (1996). Rates of erosion and sediment evacuation by glaciers: A review of field data and their implications. Global and Planetary Change, 12(1), 213–235. https://doi.org/10.1016/0921-8181(95)00021-6
66. Hambrey, M. J., & Harland, W. B. (1985). The late proterozoic glacial era. Palaeogeography, Palaeoclimatology, Palaeoecology, 51(1), 255–272. https://doi.org/10.1016/0031-0182(85)90088-4

67. Hansen, M. C., Potapov, P. V., Goetz, S. J., Turubanova, S., Tyukavina, A., Krylov, A., Kommareddy, A., & Egorov, A. (2016). Mapping tree height distributions in sub-Saharan Africa using Landsat 7 and 8 data. Remote Sens Environ, 185. https://doi.org/10.1016/j.rse.2016.02.023
68. Harper, J. T., Humphrey, N. F., & Pfeffer, W. T. (1998). Three-Dimensional Deformation Measured in an Alaskan Glacier. Science, 281(5381), 1340–1342. https://doi.org/10.1126/science.281.5381.1340
69. Hart, J. K. (1998). The deforming bed/debris-rich basal ice continuum and its implications for the formation of glacial landforms (flutes) and sediments (melt-out till). Quaternary Science Reviews, 17(8), 737–754. https://doi.org/10.1016/S0277-3791(98)00065-6
70. Hart, J. K., & Boulton, G. S. (1991). The interrelation of glaciotectonic and glaciodepositional processes within the glacial environment. Quaternary Science Reviews, 10(4), 335–350. https://doi.org/10.1016/0277-3791(91)90035-S
71. Hastenrath, S. (2009). Past glaciation in the tropics. Quaternary Science Reviews, 28(9), 790–798. https://doi.org/10.1016/j.quascirev.2008.12.004
72. Hay, W. W., Soeding, E., DeConto, R. M., & Wold, C. N. (2002). The Late Cenozoic uplift – climate change paradox. International Journal of Earth Sciences, 91(5), 746–774. https://doi.org/10.1007/s00531-002-0263-1
73. He, R., Lang, X., & Shen, B. (2021). A rapid rise of seawater δ13C during the deglaciation of the Marinoan Snowball Earth. Global and Planetary Change, 207, 103672. https://doi.org/10.1016/j.gloplacha.2021.103672
74. He, T., Liang, S., Wang, D., Cao, Y., Gao, F., Yu, Y., & Feng, M. (2018). Evaluating land surface albedo estimation from Landsat MSS, TM, ETM +, and OLI data based on the unified direct estimation approach. Remote Sensing of Environment, 204, 181–196. https://doi.org/10.1016/j.rse.2017.10.031
75. Hebbeln, D., Henrich, R., & Baumann, K.-H. (1998). PALEOCEANOGRAPHY OF THE LAST INTERGLACIAL/GLACIAL CYCLE IN THE POLAR NORTH ATLANTIC. Quaternary Science Reviews, 17(1), 125–153. https://doi.org/10.1016/S0277-3791(97)00067-X
76. Hodgson, D. A., Graham, A. G. C., Roberts, S. J., Bentley, M. J., Cofaigh, C. Ó., Verleyen, E., Vyverman, W., Jomelli, V., Favier, V., Brunstein, D., Verfaillie, D., Colhoun, E. A., Saunders, K. M., Selkirk, P. M., Mackintosh, A., Hedding, D. W., Nel, W., Hall, K., McGlone, M. S., … Smith, J. A. (2014). Terrestrial and submarine evidence for the extent and timing of the Last Glacial Maximum and the onset of deglaciation on the maritime-

Antarctic and sub-Antarctic islands. Quaternary Science Reviews, 100, 137–158. https://doi.org/10.1016/j.quascirev.2013.12.001

77. Hodson, A., Anesio, A. M., Tranter, M., Fountain, A., Osborn, M., Priscu, J., Laybourn-Parry, J., & Sattler, B. (2008). Glacial Ecosystems. Ecological Monographs, 78(1), 41–67. https://doi.org/10.1890/07-0187.1

78. Howat, I. M., Tulaczyk, S., Waddington, E., & Björnsson, H. (2008). Dynamic controls on glacier basal motion inferred from surface ice motion. Journal of Geophysical Research: Earth Surface, 113(F3). https://doi.org/10.1029/2007JF000925

79. Hughes, T. (2003). Geometrical force balance in glaciology. Journal of Geophysical Research: Solid Earth, 108(B11). https://doi.org/10.1029/2003JB002557

80. Hughes, T., Sargent, A., Fastook, J., Purdon, K., Li, J., Yan, J.-B., & Gogineni, S. (2016). Sheet, stream, and shelf flow as progressive ice-bed uncoupling: Byrd Glacier, Antarctica and Jakobshavn Isbrae, Greenland. The Cryosphere, 10(1), 193–225. https://doi.org/10.5194/tc-10-193-2016

81. Huuse, M., Lykke-Andersen, H., & Michelsen, O. (2001). Cenozoic evolution of the eastern Danish North Sea. Marine Geology, 177(3), 243–269. https://doi.org/10.1016/S0025-3227(01)00168-2

82. Imbrie, J., Berger, A., Boyle, E. A., Clemens, S. C., Duffy, A., Howard, W. R., Kukla, G., Kutzbach, J., Martinson, D. G., McIntyre, A., Mix, A. C., Molfino, B., Morley, J. J., Peterson, L. C., Pisias, N. G., Prell, W. L., Raymo, M. E., Shackleton, N. J., & Toggweiler, J. R. (1993). On the structure and origin of major glaciation cycles 2. The 100,000-year cycle. Paleoceanography, 8(6), 699–735. https://doi.org/10.1029/93PA02751

83. Katrine Elnegaard Hansen, Lorenzen, J., Davies, J., Wacker, L., Pearce, C., & Seidenkrantz, M.-S. (2022). Deglacial to Mid Holocene environmental conditions on the northeastern Greenland shelf, western Fram Strait. Quaternary Science Reviews, 293, 107704. https://doi.org/10.1016/j.quascirev.2022.107704

84. Khilchevskyi, V., & Karamushka, V. (2020). Global Water Resources: Distribution and Demand. In W. Leal Filho, A. M. Azul, L. Brandli, A. Lange Salvia, & T. Wall (Eds.), Clean Water and Sanitation (pp. 1–11). Springer International Publishing. https://doi.org/10.1007/978-3-319-70061-8_101-1

85. Kinzel, P. J., Legleiter, C. J., & Nelson, J. M. (2013). Mapping River Bathymetry With a Small Footprint Green LiDAR: Applications and Challenges1. JAWRA Journal of the American Water Resources Association, 49(1), 183–204. https://doi.org/10.1111/jawr.12008

86. Knap, W. H., & Reijmer, C. H. (1998). Anisotropy of the Reflected Radiation Field Over Melting Glacier Ice: Measurements in Landsat TM Bands 2 and 4. Remote Sensing of Environment, 65(1), 93–104. https://doi.org/10.1016/S0034-4257(98)00015-7
87. Kumar, O., Ramanathan, AL., Bakke, J., Kotlia, B. S., Shrivastava, J. P., Kumar, P., Sharma, R., & Kumar, P. (2021). Role of Indian Summer Monsoon and Westerlies on glacier variability in the Himalaya and East Africa during Late Quaternary: Review and new data. Earth-Science Reviews, 212, 103431. https://doi.org/10.1016/j.earscirev.2020.103431
88. Larsen, D. J., Miller, G. H., Geirsdóttir, Á., & Ólafsdóttir, S. (2012). Non-linear Holocene climate evolution in the North Atlantic: A high-resolution, multi-proxy record of glacier activity and environmental change from Hvítárvatn, central Iceland. Quaternary Science Reviews, 39, 14–25. https://doi.org/10.1016/j.quascirev.2012.02.006
89. Lazzara, M. A., Jezek, K. C., Scambos, T. A., MacAyeal, D. R., & Van Der Veen, C. J. (2008). On the recent calving of icebergs from the Ross ice shelf. Polar Geography, 31(1–2), 15–26. https://doi.org/10.1080/10889370802175937
90. Le Heron, D. P., Tofaif, S., & Melvin, J. (2018). Chapter 3 - The Early Palaeozoic Glacial Deposits of Gondwana: Overview, Chronology, and Controversies. In J. Menzies & J. J. M. van der Meer (Eds.), Past Glacial Environments (Second Edition) (pp. 47–73). Elsevier. https://doi.org/10.1016/B978-0-08-100524-8.00002-6
91. Lee, J. R., & Phillips, E. (2013). Glacitectonics – a key approach to examining ice dynamics, substrate rheology and ice-bed coupling. Proceedings of the Geologists' Association, 124(5), 731–737. https://doi.org/10.1016/j.pgeola.2013.07.006
92. Levy, R. H., Dolan, A. M., Escutia, C., Gasson, E. G. W., McKay, R. M., Naish, T., Patterson, M. O., Pérez, L. F., Shevenell, A. E., Flierdt, T. van de, Dickinson, W., Kowalewski, D. E., Meyers, S. R., Ohneiser, C., Sangiorgi, F., Williams, T., Chorley, H. K., Santis, L. D., Florindo, F., … Verret, M. (2022). Chapter 9—Antarctic environmental change and ice sheet evolution through the Miocene to Pliocene – a perspective from the Ross Sea and George V to Wilkes Land Coasts. In F. Florindo, M. Siegert, L. D. Santis, & T. Naish (Eds.), Antarctic Climate Evolution (Second Edition) (pp. 389–521). Elsevier. https://doi.org/10.1016/B978-0-12-819109-5.00014-1
93. Li, H., Xu, C.-Y., Beldring, S., Tallaksen, L. M., & Jain, S. K. (2016). Water Resources Under Climate Change in Himalayan Basins. Water

Resources Management, 30(2), 843–859. https://doi.org/10.1007/s11269-015-1194-5

94. Limarino, C. O., & Spalletti, L. A. (2006). Paleogeography of the upper Paleozoic basins of southern South America: An overview. Journal of South American Earth Sciences, 22(3), 134–155. https://doi.org/10.1016/j.jsames.2006.09.011

95. Loke, M. H., Chambers, J. E., Rucker, D. F., Kuras, O., & Wilkinson, P. B. (2013). Recent developments in the direct-current geoelectrical imaging method. Journal of Applied Geophysics, 95, 135–156. https://doi.org/10.1016/j.jappgeo.2013.02.017

96. M.A, J. C. (1954). The early progress of British geology.—III. Annals of Science. https://doi.org/10.1080/00033795400200174

97. Macarewich, S. I., & Poulsen, C. J. (2022). Glacial-Interglacial Controls on Ocean Circulation and Temperature During the Permo-Carboniferous. Paleoceanography and Paleoclimatology, 37(11), e2022PA004417. https://doi.org/10.1029/2022PA004417

98. Madras, G., & McCoy, B. J. (2004). Temperature effects on the transition from nucleation and growth to Ostwald ripening. Chemical Engineering Science, 59(13), 2753–2765. https://doi.org/10.1016/j.ces.2004.03.022

99. Marston, R. A. (2010). Geomorphology and vegetation on hillslopes: Interactions, dependencies, and feedback loops. Geomorphology, 116(3), 206–217. https://doi.org/10.1016/j.geomorph.2009.09.028

100. Martinerie, P., Raynaud, D., Etheridge, D. M., Barnola, J.-M., & Mazaudier, D. (1992). Physical and climatic parameters which influence the air content in polar ice. Earth and Planetary Science Letters, 112(1), 1–13. https://doi.org/10.1016/0012-821X(92)90002-D

101. Maslin, M. A., Shackleton, N. J., & Pflaumann, U. (1995). Surface water temperature, salinity, and density changes in the northeast Atlantic during the last 45,000 years: Heinrich events, deep water formation, and climatic rebounds. Paleoceanography, 10(3), 527–544. https://doi.org/10.1029/94PA03040

102. Mayewski, P. A., & Bender, M. (1995). The GISP2 ice core record—Paleoclimate highlights. Reviews of Geophysics, 33(S2), 1287–1296. https://doi.org/10.1029/95RG00498

103. Mehta, M., Kumar, V., Garg, S., & Shukla, A. (2021). Little Ice Age glacier extent and temporal changes in annual mass balance (2016–2019) of Pensilungpa Glacier, Zanskar Himalaya. Regional Environmental Change, 21(2), 38. https://doi.org/10.1007/s10113-021-01766-2

104. Menzies, J., van der Meer, J. J. M., & Shilts, W. W. (2018). Chapter 5—Subglacial Processes and Sediments. In J. Menzies & J. J. M. van der Meer (Eds.), Past Glacial Environments (Second Edition) (pp. 105–158). Elsevier. https://doi.org/10.1016/B978-0-08-100524-8.00004-X
105. Ming, J., Wang, P., Zhao, S., & Chen, P. (2013). Disturbance of light-absorbing aerosols on the albedo in a winter snowpack of Central Tibet. Journal of Environmental Sciences, 25(8), 1601–1607. https://doi.org/10.1016/S1001-0742(12)60220-4
106. Mir, R. A., Jain, S. K., Saraf, A. K., & Goswami, A. (2015). Decline in snowfall in response to temperature in Satluj basin, western Himalaya. Journal of Earth System Science, 124(2), 365–382. https://doi.org/10.1007/s12040-015-0539-z
107. Mix, A. C., Bard, E., & Schneider, R. (2001). Environmental processes of the ice age: Land, oceans, glaciers (EPILOG). Quaternary Science Reviews, 20(4), 627–657. https://doi.org/10.1016/S0277-3791(00)00145-1
108. Molden, D. J., Shrestha, A. B., Immerzeel, W. W., Maharjan, A., Rasul, G., Wester, P., Wagle, N., Pradhananga, S., & Nepal, S. (2022). The Great Glacier and Snow-Dependent Rivers of Asia and Climate Change: Heading for Troubled Waters. In A. K. Biswas & C. Tortajada (Eds.), Water Security Under Climate Change (pp. 223–250). Springer. https://doi.org/10.1007/978-981-16-5493-0_12
109. Möller, P., Larsen, N. K., Kjær, K. H., Funder, S., Schomacker, A., Linge, H., & Fabel, D. (2010). Early to middle Holocene valley glaciations on northernmost Greenland. Quaternary Science Reviews, 29(25), 3379–3398. https://doi.org/10.1016/j.quascirev.2010.06.044
110. Müller, R. D., & Dutkiewicz, A. (2018). Oceanic crustal carbon cycle drives 26-million-year atmospheric carbon dioxide periodicities. Science Advances, 4(2), eaaq0500. https://doi.org/10.1126/sciadv.aaq0500
111. Nesje, A., Bakke, J., Dahl, S. O., Lie, Ø., & Matthews, J. A. (2008). Norwegian mountain glaciers in the past, present and future. Global and Planetary Change, 60(1), 10–27. https://doi.org/10.1016/j.gloplacha.2006.08.004
112. Nisbet, E., & Fowler, C. M. R. (2011). The evolution of the atmosphere in the Archaean and early Proterozoic. Chinese Science Bulletin, 56(1), 4–13. https://doi.org/10.1007/s11434-010-4199-8
113. Nur'utami, M. N., & Hidayat, R. (2016). Influences of IOD and ENSO to Indonesian Rainfall Variability: Role of Atmosphere-ocean Interaction in the Indo-pacific Sector. Procedia Environmental Sciences, 33, 196–203. https://doi.org/10.1016/j.proenv.2016.03.070

114. Ohneiser, K., Ansmann, A., Witthuhn, J., Deneke, H., Chudnovsky, A., Walter, G., & Senf, F. (2023). Self-lofting of wildfire smoke in the troposphere and stratosphere: Simulations and space lidar observations. Atmospheric Chemistry and Physics, 23(4), 2901–2925. https://doi.org/10.5194/acp-23-2901-2023

115. Ottesen, D., Stokes, C. R., Rise, L., & Olsen, L. (2008). Ice-sheet dynamics and ice streaming along the coastal parts of northern Norway. Quaternary Science Reviews, 27(9), 922–940. https://doi.org/10.1016/j.quascirev.2008.01.014

116. Owen, L. A., & Benn, D. I. (2005). Equilibrium-line altitudes of the Last Glacial Maximum for the Himalaya and Tibet: An assessment and evaluation of results. Quaternary International, 138–139, 55–78. https://doi.org/10.1016/j.quaint.2005.02.006

117. Pages, P. I. W. G. of. (2016). Interglacials of the last 800,000 years. Reviews of Geophysics, 54(1), 162–219. https://doi.org/10.1002/2015RG000482

118. Pan, C. (2013). INVENTORY OF MONGOLIAN GLACIERS FOR THE GLOBAL LAND ICE MEASUREMENTS FROM SPACE (GLIMS) PROGRAM. Graduate Student Theses, Dissertations, & Professional Papers. https://scholarworks.umt.edu/etd/593

119. Payne, A. J., & Dongelmans, P. W. (1997). Self-organization in the thermomechanical flow of ice sheets. Journal of Geophysical Research: Solid Earth, 102(B6), 12219–12233. https://doi.org/10.1029/97JB00513

120. Peixoto, J. P. (1995). The role of the atmosphere in the water cycle. In H. R. Oliver & S. A. Oliver (Eds.), The Role of Water and the Hydrological Cycle in Global Change (pp. 199–252). Springer. https://doi.org/10.1007/978-3-642-79830-6_8

121. Pérez-Alarcón, A., Coll-Hidalgo, P., Fernández-Alvarez, J. C., Trigo, R. M., Nieto, R., & Gimeno, L. (2023). Climatological variations of moisture sources for precipitation of North Atlantic tropical cyclones linked to their tracks. Atmospheric Research, 290, 106778. https://doi.org/10.1016/j.atmosres.2023.106778

122. Petersen, S., Kuhn, K., Kuhn, T., Augustin, N., Hékinian, R., Franz, L., & Borowski, C. (2009). The geological setting of the ultramafic-hosted Logatchev hydrothermal field (14°45′N, Mid-Atlantic Ridge) and its influence on massive sulfide formation. Lithos, 112(1), 40–56. https://doi.org/10.1016/j.lithos.2009.02.008

123. Placidi, L., Hutter, K., & Faria, S. H. (2006). A Critical Review of the Mechanics of Polycrystalline Polar Ice. GAMM-Mitteilungen, 29(1), 80–117. https://doi.org/10.1002/gamm.201490025

124. Pomper, J., Salcher, B. C., Eichkitz, C., Prasicek, G., Lang, A., Lindner, M., & Götz, J. (2017). The glacially overdeepened trough of the Salzach Valley, Austria: Bedrock geometry and sedimentary fill of a major Alpine subglacial basin. Geomorphology, 295, 147–158. https://doi.org/10.1016/j.geomorph.2017.07.009

125. Rasul, G. (2014). Food, water, and energy security in South Asia: A nexus perspective from the Hindu Kush Himalayan region☆. Environmental Science & Policy, 39, 35–48. https://doi.org/10.1016/j.envsci.2014.01.010

126. Ravier, E., & Buoncristiani, J.-F. (2018). Chapter 12—Glaciohydrogeology. In J. Menzies & J. J. M. van der Meer (Eds.), Past Glacial Environments (Second Edition) (pp. 431–466). Elsevier. https://doi.org/10.1016/B978-0-08-100524-8.00013-0

127. Reeh, N. (1991). The last interglacial as recorded in the greenland ice sheet and Canadian Arctic ice caps. Quaternary International, 10–12, 123–142. https://doi.org/10.1016/1040-6182(91)90045-P

128. Rovere, A., Stocchi, P., & Vacchi, M. (2016). Eustatic and Relative Sea Level Changes. Current Climate Change Reports, 2(4), 221–231. https://doi.org/10.1007/s40641-016-0045-7

129. Ruddiman, W. F. (2003). Orbital insolation, ice volume, and greenhouse gases. Quaternary Science Reviews, 22(15), 1597–1629. https://doi.org/10.1016/S0277-3791(03)00087-8

130. Ruddiman, W. F., & Raymo, M. E. (2003). A methane-based time scale for Vostok ice. Quaternary Science Reviews, 22(2), 141–155. https://doi.org/10.1016/S0277-3791(02)00082-3

131. Schaefer, J. M., Putnam, A. E., Denton, G. H., Kaplan, M. R., Birkel, S., Doughty, A. M., Kelley, S., Barrell, D. J. A., Finkel, R. C., Winckler, G., Anderson, R. F., Ninneman, U. S., Barker, S., Schwartz, R., Andersen, B. G., & Schluechter, C. (2015). The Southern Glacial Maximum 65,000 years ago and its Unfinished Termination. Quaternary Science Reviews, 114, 52–60. https://doi.org/10.1016/j.quascirev.2015.02.009

132. Seifert, A., Leinonen, J., Siewert, C., & Kneifel, S. (2019). The Geometry of Rimed Aggregate Snowflakes: A Modeling Study. Journal of Advances in Modeling Earth Systems, 11(3), 712–731. https://doi.org/10.1029/2018MS001519

133. Seligman, G. (1949). Research on Glacier Flow. Geografiska Annaler. https://www.tandfonline.com/doi/abs/10.1080/20014422.1949.11880808

134. Shen, C.-C., Lee, T., Chen, C.-Y., Wang, C.-H., Dai, C.-F., & Li, L.-A. (1996). The calibration of D[Sr/Ca]versus sea surface temperature relationship for Porites corals. Geochimica et Cosmochimica Acta, 60(20), 3849–3858. https://doi.org/10.1016/0016-7037(96)00205-0

135. Siebe, C., Salinas, S., Arana-Salinas, L., Macías, J. L., Gardner, J., & Bonasia, R. (2017). The ~ 23,500 y 14C BP White Pumice Plinian eruption and associated debris avalanche and Tochimilco lava flow of Popocatépetl volcano, México. Journal of Volcanology and Geothermal Research, 333–334, 66–95. https://doi.org/10.1016/j.jvolgeores.2017.01.011

136. Siegert, M. J. (2000). Antarctic subglacial lakes. Earth-Science Reviews, 50(1), 29–50. https://doi.org/10.1016/S0012-8252(99)00068-9

137. Sonnek, K. M., Mårtensson, T., Veibäck, E., Tunved, P., Grahn, H., von Schoenberg, P., Brännström, N., & Bucht, A. (2017). The impacts of a Laki-like eruption on the present Swedish society. Natural Hazards, 88(3), 1565–1590. https://doi.org/10.1007/s11069-017-2933-0

138. Souchez, R. (1997). The buildup of the ice sheet in central Greenland. Journal of Geophysical Research: Oceans, 102(C12), 26317–26323. https://doi.org/10.1029/96JC01558

139. Steinthorsdottir, M., Coxall, H. K., de Boer, A. M., Huber, M., Barbolini, N., Bradshaw, C. D., Burls, N. J., Feakins, S. J., Gasson, E., Henderiks, J., Holbourn, A. E., Kiel, S., Kohn, M. J., Knorr, G., Kürschner, W. M., Lear, C. H., Liebrand, D., Lunt, D. J., Mörs, T., … Strömberg, C. a. E. (2021). The Miocene: The Future of the Past. Paleoceanography and Paleoclimatology, 36(4), e2020PA004037. https://doi.org/10.1029/2020PA004037

140. Su, F., Zhang, L., Ou, T., Chen, D., Yao, T., Tong, K., & Qi, Y. (2016). Hydrological response to future climate changes for the major upstream river basins in the Tibetan Plateau. Global and Planetary Change, 136, 82–95. https://doi.org/10.1016/j.gloplacha.2015.10.012

141. Swanson-Hysell, N. L. (2021). Chapter 4—The Precambrian paleogeography of Laurentia. In L. J. Pesonen, J. Salminen, S.-Å. Elming, D. A. D. Evans, & T. Veikkolainen (Eds.), Ancient Supercontinents and the Paleogeography of Earth (pp. 109–153). Elsevier. https://doi.org/10.1016/B978-0-12-818533-9.00009-6

142. Syvitski, J. P. M. (1989). On the deposition of sediment within glacier-influenced fjords: Oceanographic controls. Marine Geology, 85(2), 301–329. https://doi.org/10.1016/0025-3227(89)90158-8

143. Tang, Y., Han, Y., Ma, X., & Liu, Z. (2018). Elevated heat pump effects of dust aerosol over Northwestern China during summer. Atmospheric Research, 203, 95–104. https://doi.org/10.1016/j.atmosres.2017.12.004

144. Telbisz, T., Tóth, G., A. Ruban, D., & M. Gutak, J. (2019). Notable Glaciokarsts of the World. In M. Veress, T. Telbisz, G. Tóth, D. Lóczy, D. A. Ruban, & J. M. Gutak (Eds.), Glaciokarsts (pp. 373–499). Springer International Publishing. https://doi.org/10.1007/978-3-319-97292-3_9

145. Todhunter, P. E. (2021). Hydrological basis of the Devils Lake, North Dakota (USA), terminal lake flood disaster. Natural Hazards, 106(3), 2797–2824. https://doi.org/10.1007/s11069-021-04567-2
146. Tran, C. D. (2003). Infrared Multispectral Imaging: Principles and Instrumentation. Applied Spectroscopy Reviews, 38(2), 133–153. https://doi.org/10.1081/ASR-120021165
147. Tripathee, L., Gul, C., Kang, S., Chen, P., Huang, J., & Rai, M. (2021). Transport Mechanisms, Potential Sources, and Radiative Impacts of Black Carbon Aerosols on the Himalayas and Tibetan Plateau Glaciers. In S. Tiwari & P. Saxena (Eds.), Air Pollution and Its Complications: From the Regional to the Global Scale (pp. 7–23). Springer International Publishing. https://doi.org/10.1007/978-3-030-70509-1_2
148. van Angelen, J. H., van den Broeke, M. R., Wouters, B., & Lenaerts, J. T. M. (2014). Contemporary (1960–2012) Evolution of the Climate and Surface Mass Balance of the Greenland Ice Sheet. Surveys in Geophysics, 35(5), 1155–1174. https://doi.org/10.1007/s10712-013-9261-z
149. Van Wychen, W., Copland, L., & Burgess, D. (2020). Ice Masses of the Eastern Canadian Arctic Archipelago. In O. Slaymaker & N. Catto (Eds.), Landscapes and Landforms of Eastern Canada (pp. 297–314). Springer International Publishing. https://doi.org/10.1007/978-3-030-35137-3_13
150. Vaughan, D. G. (2008). West Antarctic Ice Sheet collapse – the fall and rise of a paradigm. Climatic Change, 91(1), 65–79. https://doi.org/10.1007/s10584-008-9448-3
151. Wakeford, R., Darby, S. C., & Murphy, M. F. G. (2010). Temporal trends in childhood leukaemia incidence following exposure to radioactive fallout from atmospheric nuclear weapons testing. Radiation and Environmental Biophysics, 49(2), 213–227. https://doi.org/10.1007/s00411-010-0266-4
152. Waller, R. I., Murton, J. B., & Kristensen, L. (2012). Glacier–permafrost interactions: Processes, products and glaciological implications. Sedimentary Geology, 255–256, 1–28. https://doi.org/10.1016/j.sedgeo.2012.02.005
153. Wang, C., Deser, C., Yu, J.-Y., DiNezio, P., & Clement, A. (2017). El Niño and Southern Oscillation (ENSO): A Review. In P. W. Glynn, D. P. Manzello, & I. C. Enochs (Eds.), Coral Reefs of the Eastern Tropical Pacific: Persistence and Loss in a Dynamic Environment (pp. 85–106). Springer Netherlands. https://doi.org/10.1007/978-94-017-7499-4_4
154. Wang, X., Cheng, X., Gong, P., Huang, H., Li, Z., & Li, X. (2011). Earth science applications of ICESat/GLAS: A review. International Journal of Remote Sensing, 32(23), 8837–8864. https://doi.org/10.1080/01431161.2010.547533

155. Wani, J. M., Thayyen, R. J., Gruber, S., Ojha, C. S. P., & Stumm, D. (2020). Single-year thermal regime and inferred permafrost occurrence in the upper Ganglass catchment of the cold-arid Himalaya, Ladakh, India. Science of The Total Environment, 703, 134631. https://doi.org/10.1016/j.scitotenv.2019.134631

156. Warren, C. R. (1993). Rapid Recent Fluctuations of the Calving San Rafael Glacier, Chilean Patagonia: Climatic or Non-Climatic? Geografiska Annaler: Series A, Physical Geography, 75(3), 111–125. https://doi.org/10.1080/04353676.1993.11880389

157. Weertman, J. (1972). General theory of water flow at the base of a glacier or ice sheet. Reviews of Geophysics, 10(1), 287–333. https://doi.org/10.1029/RG010i001p00287

158. Wiersma, A. P., & Renssen, H. (2006). Model–data comparison for the 8.2 ka BP event: Confirmation of a forcing mechanism by catastrophic drainage of Laurentide Lakes. Quaternary Science Reviews, 25(1), 63–88. https://doi.org/10.1016/j.quascirev.2005.07.009

159. Willeit, M., Ganopolski, A., Calov, R., Robinson, A., & Maslin, M. (2015). The role of CO2 decline for the onset of Northern Hemisphere glaciation. Quaternary Science Reviews, 119, 22–34. https://doi.org/10.1016/j.quascirev.2015.04.015

160. Wolff, E. W., Barbante, C., Becagli, S., Bigler, M., Boutron, C. F., Castellano, E., de Angelis, M., Federer, U., Fischer, H., Fundel, F., Hansson, M., Hutterli, M., Jonsell, U., Karlin, T., Kaufmann, P., Lambert, F., Littot, G. C., Mulvaney, R., Röthlisberger, R., … Wegner, A. (2010). Changes in environment over the last 800,000 years from chemical analysis of the EPICA Dome C ice core. Quaternary Science Reviews, 29(1), 285–295. https://doi.org/10.1016/j.quascirev.2009.06.013

161. Wolken, G. J., Sharp, M., & Wang, L. (2009). Snow and ice facies variability and ice layer formation on Canadian Arctic ice caps, 1999–2005. Journal of Geophysical Research: Earth Surface, 114(F3). https://doi.org/10.1029/2008JF001173

162. Yadav, R. R., Misra, K. G., Yadava, A. K., Kotlia, B. S., & Misra, S. (2015). Tree-ring footprints of drought variability in last ~300 years over Kumaun Himalaya, India and its relationship with crop productivity. Quaternary Science Reviews, 117, 113–123. https://doi.org/10.1016/j.quascirev.2015.04.003

163. You, Q., Chen, D., Wu, F., Pepin, N., Cai, Z., Ahrens, B., Jiang, Z., Wu, Z., Kang, S., & AghaKouchak, A. (2020). Elevation dependent warming over the Tibetan Plateau: Patterns, mechanisms and perspectives. Earth-Science Reviews, 210, 103349. https://doi.org/10.1016/j.earscirev.2020.103349

164. Young, G. M. (2019). Aspects of the Archean-Proterozoic transition: How the great Huronian Glacial Event was initiated by rift-related uplift and terminated at the rift-drift transition during break-up of Lauroscandia. Earth-Science Reviews, 190, 171–189. https://doi.org/10.1016/j.earscirev.2018.12.013

165. Young, G. M., & Nesbitt, H. W. (1985). The Gowganda Formation in the southern part of the Huronian outcrop belt, Ontario, Canada: Stratigraphy, depositional environments and regional tectonic significance. Precambrian Research, 29(1), 265–301. https://doi.org/10.1016/0301-9268(85)90071-3

166. Young, N. E., Briner, J. P., Miller, G. H., Lesnek, A. J., Crump, S. E., Thomas, E. K., Pendleton, S. L., Cuzzone, J., Lamp, J., Zimmerman, S., Caffee, M., & Schaefer, J. M. (2020). Deglaciation of the Greenland and Laurentide ice sheets interrupted by glacier advance during abrupt coolings. Quaternary Science Reviews, 229, 106091. https://doi.org/10.1016/j.quascirev.2019.106091

167. Yu, W., Algeo, T. J., Zhou, Q., Du, Y., & Wang, P. (2020). Cryogenian cap carbonate models: A review and critical assessment. Palaeogeography, Palaeoclimatology, Palaeoecology, 552, 109727. https://doi.org/10.1016/j.palaeo.2020.109727

168. Zhou, S., Nakawo, M., Hashimoto, S., & Sakai, A. (2008). The effect of refreezing on the isotopic composition of melting snowpack. Hydrological Processes, 22(6), 873–882. https://doi.org/10.1002/hyp.6662

INDEX

C

D

E

H

I

J

L

M

FSC
www.fsc.org
MIX
Papier | Fördert gute Waldnutzung
FSC® C083411

Kolibri 360 GmbH
Ferdinand-Jühlke-Straße 7
99095 Erfurt, Deutschland
produktsicherheit@kolibri360.de